The Consultant's
AI Companion

◆

Everything You Need to Succeed in the Age of Artificial Intelligence

ROB BERG

súpren
PUBLISHING

Paperback ISBN-13: 979-8-9930673-0-8
eBook ISBN-13: 979-8-9930673-1-5
Library of Congress Control Number: 2025920575
Designed by Rob Berg

SUPREN PUBLISHING
7643 GATE PKWY, STE 104
JACKSONVILLE, FL 32256
www.suprenpublishing.com

For Christine, Jasper, and Rupert

Contents

PART 2: TOOLS & TACTICS

Chapter 4: Out-of-the-Box AI Copilots for Consultants.................... 57

Chapter 5: Building Custom AI Assistants..................................... 79

CONTENTS

PART 3: PLAYBOOKS & PROMPTS

CONTENTS

CONTENTS

Preface

It's here. The Age of Artificial Intelligence has arrived. On November 30, 2022, life as we knew it changed forever. That was the day that ChatGPT was released to the public. And the blizzard of prognostications and hype and fear and wonder has been swirling out of control ever since.

Of course, AI is not a new thing. Its origins stretch back for decades. Indeed, I recall many discussions with a former roommate in the late 1980s about his work in AI (expert systems at the time) at a defense contractor on Long Island. And AI's modern history predates that by another forty years.

But it was the public release of ChatGPT that truly awakened the world to AI's potential. Has it been overhyped? Without a doubt. Have we tapped even a fraction of what it can do? Not even close. Remember Bill Gates' famous observation, "We always overestimate the change that will occur in the next two years and underestimate the change that will occur in the next ten." Often referred to as Gates' Law (though many others have expressed similar sentiments), it's more relevant than ever. If we take that pivotal day in November 2022 as our starting line, the road to realizing the hype is still long—but the window to get there is closing. But make no mistake—we *will* get there.

This book, I must point out, was not motivated by hype. It was motivated by pragmatism—by my own use of AI to support my work as a consultant and my desire to help others obtain the same remarkable benefits. I've been genuinely blown away at how valuable it's become in my daily routine. The efficiency gains have been staggering:

- Proposals completed in minutes instead of hours or days.

- White papers that used to take a month or more to draft, edit, and publish now completed in just two days.

- Contract analyses performed with a few carefully selected words to prompt an LLM rather than engaging outside counsel for superficial (and expensive) legal reviews.

- Coherent frameworks to support a variety of consulting engagement types created in a few minutes.

- Company backgrounders and executive profiles in advance of high-stakes meetings with clients and prospects delivered in mere seconds.

- Beautifully crafted slide presentations created in twenty minutes or less (with initial drafts produced in as little as twenty seconds).

- Summarizations of dense subject matter that reduce thirty-page academic papers to a few pertinent points in less than a minute.

- Real-time transcriptions of meetings that are instantly summarized and delivered to meeting participants with key takeaways and action items minutes after the meeting concludes.

- An always-available, expert sounding board, capable of role-playing client positions or critiquing engagement approaches, deliverables, and client communications.

- A co-author and editor for this book that helped me to get done in four months what would ordinarily take up to a year.

That's right. This book was crafted with the assistance of ChatGPT-4o and ChatGPT's Deep Research capability. And when I say "with the assistance of," that's precisely what I mean. Because without my intervention this would largely be a work of fiction, riddled with jargon, repetitive phrases, words, ideas—and errors. It took a genuinely collaborative effort to pull this thing together. And here's how I (we?) did it:

I began by creating a "project" in ChatGPT and uploading an outline of topics—each, a chapter—to the project's "project files" area.

Next, I uploaded some of my prior writings, including the full text of my last book, *The Courageous Consultant*, white papers I had written, and transcripts from ten video lessons from a consulting course I had developed. These provided guidance to the LLM on my writing style to minimize generic AI-generated drivel. I also created and uploaded a style guide, which included my preferences for tone, style, voice, sentence length, etc., and an alignment memo detailing my core philosophies (e.g., human-centered consulting, differentiation through authenticity, coaching as a consulting superpower, the courage to forge your own path, etc.) to ensure the book reflected those sensibilities. It was only after those artifacts were created, vetted, and uploaded that I began drafting.

Drafting the copy started with an expansion of the outline. When prompted, ChatGPT suggested a few additions, added bullet points for each chapter, and offered an improved way to organize the chapters into larger categories that became Parts 1, 2, and 3 of this book.

For each chapter, I prompted ChatGPT to create a narrative, including verifiable anecdotes and, where appropriate, descriptions of relevant AI tools to support each chapter's examples. I then prompted ChatGPT to expand each chapter's bullet points, which enhanced the initial text. After reviewing that initial work product in depth, I prompted the AI to further expand the copy, which I again reviewed and edited. This usually involved multiple iterations and was done successively for each of the fifteen chapters in the book (I would leave this preface, the afterword, and appendices for later).

For the next step, I used ChatGPT's Deep Research feature. I prompted the AI to expand even further on each chapter and include well-researched and properly referenced information. The output from this step, which took from ten to thirty minutes per chapter depending on the subject matter, was the final AI-produced work product for each of the fifteen chapters.

Next came the heavy lifting for me: I read through and edited each chapter, seeking to refine the copy to ensure it reflected my desired writing style and content. For quotes, statistics, and references, I ensured they were accurate and tied to credible, verifiable sources. (In many cases, even ChatGPT's Deep Research returned utter nonsense and links to references that simply did not exist.)

Finally, I had a far greater hand in producing this preface as I wrote the initial draft and had ChatGPT suggest (not make) edits, some of which I accepted, and many of which I did not.

A Note on Structure and Repetition

This book is designed as a reference guide, not necessarily a linear read. Each chapter is meant to stand on its own, addressing distinct aspects of AI adoption in consulting. As a result, you may notice occasional overlaps between chapters, particularly when it comes to tools or frameworks that are applicable across multiple scenarios. This is intentional.

Rather than require you to read prior chapters to understand a concept, I've chosen to reintroduce key ideas where they're most relevant. If you read the book front to back, some repetition will be apparent. But if you're using it as I hope you will—as a practical guide to return to again and again—I trust you'll find the modular format both efficient and accessible.

Book Development

I drafted an initial outline for this book on May 10, 2025. I thereafter wrote this preface, after generating and reviewing the raw text, on June 1, 2025, at which point the book's pre-edited word count exceeded 100,000. As I approach the completed product—after a lengthy period of editing, formatting, and otherwise preparing for publication—it's September 7, 2025. That's less than *four months* from concept to finished product working evenings, weekends, and a total of about ten days of scheduled time off from my consulting work. For comparison, my last book ran about 50,000 words, took roughly six months to write (with the help of a paid writing coach), and another four before it was published and available for sale. So, all told, I saved about six months of effort and *thousands* of dollars in coaching, creative, and publishing fees to produce a book that's twice as long and arguably more practical for the reader. Yet another testament to the power of AI.

As I worked through the process, a profound insight occurred to me: I learned far more by taking this approach than I might have by writing without the help of AI. As ChatGPT cranked out outlines and narratives, new ideas emerged that I hadn't previously thought of. Those ideas sparked my curiosity in a way that led me down new paths that yielded new knowledge and new insights. I became familiar with a multitude of AI tools and methods because of ChatGPT's outputs, rather than simply limiting the scope to what I had already known or understood. AI-generated responses caused me to pause and consider conflicting perspectives and dive deeply into areas I might not have ordinarily cared about. And this is the magic—and *it is magic*—of AI as a collaborative partner, as a working companion. You're cheating yourself if you let AI do all the work for you. With AI, you have the benefit of an expert colleague, always available to answer questions, always willing to challenge your ideas in a way that helps you to improve. Take advantage of that and your own capabilities will expand dramatically, your ability to respond will become supercharged, and your value will be enhanced immeasurably.

This final product is a determined attempt to share with my fellow consultants the many wonders of AI so they, too, can gain extraordinary efficiencies and enhance the value of their work with far less effort than even thought possible just a few years ago. I also set out to prove my thesis that AI is, indeed, worthy of the hype.

I do hope you find genuine value in what follows and, more importantly, elevate your consulting game by using AI in a collaborative way—a way that allows you to focus your attention on those uniquely human qualities, like creative problem solving and critical thinking, that are the immutable hallmarks of exceptional advisors.

Rob Berg
St. Augustine, Florida
September 2025

ROB BERG

Part 1

Mindset & Strategy

Before we dive into practical tools, workflows, or use cases, let's set a mood: Part 1 is about mindset. As consultants, in addition to adopting AI, we need to decide what kind of professionals we want to be in an AI-driven world. This section helps set the strategic and philosophical foundation for everything that follows. It challenges the reader to think beyond the hype and ask, "What will remain distinctly human in my practice?" and "Where can AI amplify—not replace—what I do best?"

In these first three chapters, we'll explore the promise and peril of AI for consultants, outline the mindset required to navigate this evolving landscape, and offer a blueprint for strategically integrating AI into your practice. Our goal isn't to create AI experts—it's to empower thoughtful consultants to become more creative, more intentional, and more valuable by understanding the leverage AI makes possible. Think of this part as your compass, orienting you not toward a fixed destination, but toward a way of thinking that will keep you relevant, courageous, and client-centric—no matter how fast the terrain changes.

CHAPTER ONE

◆

The Promise and Peril of AI for Consultants

et's be clear from the outset: AI is not coming for your job. But it is coming for your habits. And that's exactly where the opportunity lies. For consultants who pride themselves on being strategic thinkers, trusted advisors, and catalysts for client transformation, the emergence of generative AI is both a wake-up call and a gift. It challenges assumptions about what constitutes value in our work—and, if approached with curiosity and discernment, reveals entirely new ways to deepen our impact.

This chapter will help you to develop a clear-eyed perspective on what AI means for the future of consulting in terms of tools and automation as well as how we show up for our clients, how we differentiate ourselves, and how we choose to evolve.

What's Changing—and What's Not

AI is reshaping knowledge work in an unprecedented manner. Tasks once reserved for seasoned analysts, researchers, and PowerPoint experts can now be completed in seconds by large language models trained on vast troves of data. And tasks involving data collection, synthesis, and basic reporting are among the most susceptible to automation across professional services. Similarly, while AI excels at speed, data processing, and pattern recognition, it falters when it comes to nuanced judgment and context. It struggles with creative problem-solving and managing complex stakeholder dynamics— capabilities central to high-value consulting. In short, what's changing is *how* certain tasks get done, but not the core reasons *why* clients seek out consultants.

So those overhyped predictions that warn of consulting's imminent demise are, to borrow from Mark Twain, greatly exaggerated. In fact, our uniquely human strengths are becoming *more* valuable precisely because they can't be easily codified. Clients don't hire consultants for perfectly worded reports or technically flawless models; they hire us for our judgment. For discernment. For the ability to navigate ambiguity, to see patterns others don't, and to speak truth to power when no one else will. AI can support these abilities, but it can't replicate them. The fundamental human elements of consulting remain unchanged: the trust we build, the courage we model, and the empathy and creativity we bring—enduring qualities that are genuine differentiators in an AI-saturated world.

The Myth of AI Omnipotence

Recent narratives suggest AI will soon replace human expertise entirely, yet history tells a different story. In every major technological shift over the past few decades, whether it was the rise of ERP systems in the 1990s, the internet boom in the 2000s, or the cloud computing era of the 2010s, automation may have changed what consultants do, but it also increased demand for what only humans could provide. For example, when enterprise software began to proliferate, veritable armies of consultants were mobilized to customize and implement complex systems, fueling growth in IT and strategy consulting. The commercialization of the internet initially made information more accessible and formerly obscure products and services instantly available to the masses while introducing a bewildering array of new strategic questions for which savvy clients turned to consultants for guidance. And as cloud computing reduced IT infrastructure burdens, businesses sought consultants to help navigate the challenges of migration, integration, and innovation in the cloud. The lesson is clear; tools evolve, but human value endures. Each technological leap freed consultants from grunt work (like manual data crunching or paper-based research) and *amplified* the importance of higher-order skills like strategic thinking, creative vision, and trusted advisor relationships. Far from rendering the consultant irrelevant, past tech revolutions ultimately refocused consultants on the uniquely

4

human aspects of their role. AI is no different. It will undoubtedly automate many tasks we do today, just as past technologies have, but it will also open new fronts where clients urgently need our insight. The promise and peril of AI for consultants echo those of earlier eras. Those who adapt and embrace new tools will thrive, while those clinging to old models will struggle. The smart move, as history shows, is to leverage technology to augment our human strengths, not to assume it might negate them.

The Amplifier Effect

Think of AI as an amplifier. If you're an empathic listener, AI can help you detect patterns in client feedback or survey data that deepen your understanding. If you're a systems thinker, AI can reveal connections across complex workflows that better inform your recommendations. If you're a creative problem solver, AI can serve as a tireless brainstorming partner, generating a plethora of new ideas for you to entertain, investigate, and implement. In these and many other ways to be discussed on the following pages, AI can dramatically extend the reach of your natural talents.

But amplification cuts both ways. If you're vague, AI will generate vague outputs. If your thinking is muddled, it will multiply the confusion. In other words, AI doesn't fix bad consulting; it makes it louder. This "garbage in, garbage out" dynamic means that improving our own clarity and critical thinking is more important than ever. As leading AI expert Andrew Ng has noted, "AI won't replace human workers, but people who use it will replace those who don't." The key is using it wisely.

AI's Real Impact on Research Tasks

A Harvard Business School field experiment with Boston Consulting Group in 2023 found that consultants using GPT-4 completed tasks 25 percent faster and produced over 40 percent higher-quality results compared to those without AI assistance. Interestingly, those who were less experienced benefited the most, narrowing performance gaps. This study suggests that AI can significantly enhance productivity in consulting tasks, especially by speeding up research, analysis, and drafting work, thereby freeing consultants to spend more time on higher-order thinking. It's a striking confirmation that AI, used judiciously, serves as a force multiplier for our efforts, an *amplifier* of efficiency and quality rather than a replacement. This underscores AI's role as an amplifier, not a substitute; it can elevate the floor of our work, but the ceiling is still set by human insight.

This is why self-awareness is more essential than ever. Knowing your strengths, your blind spots, and your unique consulting style is key to choosing and using AI tools in a way that enhances—rather than erodes—your unique value. In the AI era, the mirror we must hold up is to ourselves: Are we using these tools to augment our expertise or to avoid the hard work of truly thinking? Honest reflection on this question separates those who simply produce more noise with AI from those who produce greater value.

The Amplifier Equation

Value Delivered = (Consultant's Insight + AI Efficiency)

× Client Relevance

This simple formula reminds us that AI's speed and capacity only multiply the consultant's underlying insight—and only to the extent that the work is relevant to the client's real needs. If you have shallow insight or pursue the wrong problem, AI will just get you to the wrong answer faster. If you solve the right problem and apply deep expertise, AI can help you deliver that solution with unprecedented efficiency. The multiplication by client relevance is a critical kicker: no matter how clever the tech, if the outcome doesn't matter to the client's objectives, it misses the mark. In short, focus on what matters and then supercharge it with AI.

Avoiding the AI Gimmick Trap

In every era of consulting, there's a temptation to chase the next shiny new object. In the early 2000s, it was Six Sigma (full disclosure: I am a recovering Six Sigma Black Belt). Later, Agile (including, but not limited to Agile development, project management, organizations, marketing, HR, manufacturing, etc.). Then digital transformation (which I liken to balling your fingers into a fist, which is, of course, a different sort of digital transformation but valid nonetheless). Today, it's AI. The challenge is similar, in that consultants might adopt a hot new trend not because it truly aligns with their practice or benefits their clients, but because they fear being left behind. The result? Over-promising, under-delivering, and a proliferation of one-size-fits-all "solutions" that do little to solve real problems while doing little more than preserving the consultant's short-term cash flow.

The Over-Automation Pitfall

Here's a completely made-up story that illustrates this problem nicely (and likely resonates with those of you who've delved into the underworld of AI consulting). Innovare Partners is a mid-sized strategy consultancy that fell into this trap. In 2024, sensing the buzz around AI, Innovare hastily rolled out a new offering branded "AI-Powered Transformation." They invested in flashy marketing and promised clients cutting-edge AI-driven insights on any problem they could conjure. But behind the scenes, Innovare's team had little AI expertise. They hadn't upskilled their staff or thought through how AI fit within their existing consulting frameworks. In one engagement, they deployed an AI tool to generate a market analysis for a client—without human vetting—and delivered the report as-is. The client quickly realized the analysis was riddled with irrelevant data and overlooked industry nuances. It felt automated and generic. Trust evaporated. Within weeks, that flagship client pulled out of the project, and a public statement from the client's CEO cited "a lack of human insight" in Innovare's approach. The fiasco was a blow to Innovare's reputation. It became clear that in their rush to appear innovative, they had treated AI as a gimmick—a shiny add-on—rather than integrating it thoughtfully. Meanwhile, a competitor, Axis Advisory, took a more cautious route. For six months, Axis quietly upskilled its consultants in data science basics and partnered with a boutique AI firm to co-develop solutions. When Axis finally launched its AI-driven service, they targeted it narrowly at customer churn analysis, an area where they had deep domain knowledge plus new AI tools. The result was a tailored service line that impressed clients and boosted revenues. Axis's clients raved that the consultants clearly understood their business and just used AI to augment that understanding, not replace it. Innovare's tale is fictional, but it's representative of a very real dynamic: using AI without purpose can backfire, while using AI with humility and alignment can deliver true value.

Gimmick or Growth?

While AI tools are becoming increasingly sophisticated, clients can often tell when AI is being used as a gimmick rather than a genuine value-add. It's a sort of an *Uncanny Valley* of consulting. (For the uninitiated, the Uncanny Valley describes the unsettling feeling evoked by humanlike figures that are almost, but not quite, perfectly realistic.) Authenticity and relevance remain the differentiators. The antidote is simple (but not always easy), and it involves grounding your use of AI in empathy and context. Ask yourself: "Does this tool really help me understand my client better?" "Does it help my client make a more courageous decision?" "Does it reinforce the relationship, or just replace something we already do?" If you can't answer these with confidence, you might be chasing a gimmick. Used wisely, on the other hand, AI can free us from low-value busywork and meaningfully increase the time and attention we have for the aspects of consulting clients value most. The goal is to augment, not alienate. A client who sees you using AI in a way that clearly advances their goals will welcome it; a client who senses you're using them as a testing ground for trendy tech will walk.

Put simply, don't let the technology eclipse the human touch. We must be intentional in our quest to ensure AI plays a supporting role, not command center stage. It's there to enable better client service, not to be a talking point in a sales pitch. Maintaining this perspective keeps us honest and keeps the client's true needs front and center.

A Moment of Reckoning and Renewal

Consultants who deliberately hoard information, churn out templated deliverables, or espouse linear, pedantic problem-solving frameworks are in for a rough ride. Aspects of consulting that can be automated *will* be automated. Recent insights shared by industry analysts underscore this inflection point. Clients increasingly expect their consultants to keep pace with technology by using AI thoughtfully and ethically to produce better outcomes. In other words, most clients now assume their estimable advisors will leverage AI to work smarter. They also demand transparency; consulting firms themselves consider transparency around AI usage to be crucial, since clients want to know how AI is being used in their projects. The message is clear; if you're not embracing these tools and explaining how you're using them, you risk losing client trust.

This might feel like a reckoning for those of us used to the old playbooks. But for consultants who view consulting as a craft—a human-centered act of diagnosis, insight, and co-creation—this moment is actually one of re-newal. Why? Because as the mechanical parts of our work get automated, we can finally double down on those aspects of consulting that made us fall in love with this profession in the first place. We get to spend more time with clients, tackling ambiguous problems, crafting creative strategies, and coaching leaders through change. It's a chance to reclaim the high ground of consulting.

Why Clients Still Pay for Judgment

It's almost axiomatic to state that clients value their consultants' ability to make solid judgments under uncertain conditions—being the cooler heads that prevail when unfettered change and its cousin, constant chaos, disturb their business-as-usual. This idea speaks volumes. For while AI can churn out endless insights, analyses, even recommendations, it can't weigh the messy human factors that drive decisions in the real world. It doesn't know how to factor in organizational politics, market timing, or cultural nuance. It can't walk into a boardroom and read the room of skeptical executives. It can't look a CEO in the eye and say, "I hear your question, but I think you're solving the wrong problem." It can't hold space for a client's anxiety about a risky decision, or intuit when to push vs. when to step back. It can't do any of that. But you can. And clients know it. That's why they continue to pay us. Indeed, the future of our industry will increasingly hinge upon relation-ships that are built on uniquely human traits like judgment, morals, ethics, and integrity in ways that are utterly inaccessible to technology. AI can pro-duce answers, but wisdom remains a human domain.

And that's why your work still matters—not *despite* AI, but in many ways *because of it*. The more AI proliferates, the more it will highlight the contrast between generic output and genuine expertise. Your human judgment be-comes not a nice-to-have, but the deciding factor. This is our opportunity to lean into that reality. Rather than compete with the machine on what it does well, we double down on what *we* do well. We sharpen our ability to formulate the right problems, to ask the provocative questions, to connect dots that aren't obvious, and to deliver advice with empathy and ethical clarity.

This book is your companion in navigating that paradox. It wasn't written to teach you how to write Python code or fine tune a large language model. There are plenty of manuals for that. Instead, our focus is on helping you become an AI-enhanced version of the consultant you already are (or aspire to be). It's about integrating these powerful tools into your practice in a way that aligns with your values and amplifies your impact. The underlying premise is that the future doesn't belong to the merely technical; it belongs to the *intentional*. Consultants who approach AI (and every new tool) with intention—who are deliberate about how they use it, why they use it, and when *not* to use it—will build deeper trust with clients and deliver better results than those who simply let the tech run roughshod.

The Consultant's Value Pyramid in the AI Era

To make this discussion concrete, consider a simple model we'll call the *Consultant's Value Pyramid*. Imagine a pyramid with three layers of consulting work:

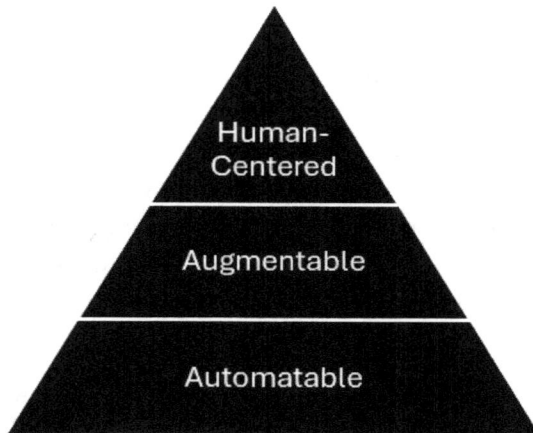

The Consultant's Value Pyramid

- **Base Layer: Automatable.** This is the foundation of basic tasks that AI and automation can largely handle now. It includes things like data gathering, research, benchmarking studies, compiling reports, and first-draft writing. These activities, while essential as inputs, no longer differentiate a consultant; they can be done faster and cheaper by AI. If most of your time is spent here, you're operating in commodity territory. In the AI era, we should aim to minimize the time we personally spend on base-layer work by delegating it to tools or junior staff, because it's not where we deliver unique value.

- **Middle Layer: Augmentable.** The middle of the pyramid is where AI can assist but not fully replace. This includes tasks like deeper analysis, insight generation, scenario planning, modeling options, and otherwise crunching data. AI can turbocharge these activities, for example by generating simulations or highlighting patterns. But it still requires a human consultant to guide it and interpret the results. Work in this layer benefits from human–AI collaboration. For instance, you might use an AI tool to iterate through dozens of strategic scenarios, but it takes your consulting experience to identify which scenarios are truly viable or relevant to the client. Consultants should strive to push more of their work into this augmentable layer, using AI to stretch their analytical capacity while still leading with human judgment.

- **Top Layer: Human-Centered.** At the top of the pyramid are the roles and outputs that only talented human consultants can provide. This is the realm of complex problem framing, judgment under uncertainty, empathic communication, facilitation of tough discussions, and "courageous advising," i.e., telling clients the truth they need to hear, in a way they can hear it. It's also where coaching comes in, as we help clients to develop their own capabilities by tapping into their unique experiences and knowledge. AI cannot replicate genuine leadership coaching, creative vision, or trust-building. Work at this level is highly context-dependent and requires emotional intelligence, ethics, and experience. This is where consultants truly differentiate themselves and deliver transformative value.

Using the Value Pyramid for Growth

How can you use the Value Pyramid as a tool in your career? First, use it for self-assessment. Take an honest look at your recent projects and break down your activities. How much of your time is spent in base-layer tasks vs. higher layers? If you realize, for example, that 70 percent of your time is spent cranking out research summaries or routine analyses, that's a signal to up-level your game (or leverage AI to handle those parts). Set a goal to delegate or automate lower-layer work so you can concentrate on middle- and top-layer contributions. This might mean learning new AI tools to handle your data prep, giving you more time to brainstorm novel insights and craft compelling narratives for your clients. Second, use the pyramid as a marketing tool. When engaging with prospective clients (or pitching new projects to existing ones), explicitly articulate where you'll be focusing your effort. For example, "Our team will use advanced AI research tools to gather the market data in days instead of weeks, allowing us to spend most of our time with you on scenario planning and decision-making—the things that really matter." By doing so, you reassure clients that you're not wasting their budget on tasks a machine could do, and you highlight the premium, human judgment you're bringing. It's a great way to differentiate your services in an AI-enhanced world. Consultants who effectively communicate that they operate at the top of the value pyramid (while smartly leveraging tech at the base) will stand out as strategic partners, not interchangeable problem-solvers.

In the end, the promise of AI for consultants is that it can liberate us from the grind and enable us to spend more time in our "zone of genius." The peril is that, if misused, it can tempt us to cut corners or homogenize our work. The balance we strike will determine the kind of consultants we become in this new era. Will we use AI to become bolder, more courageous advisors, or merely faster presentation designers? The choice, thankfully, remains ours. As we continue through this book, we'll explore how to harness AI in ways that expand your capacity without compromising your humanity. The path of the AI-enhanced consultant is not about human vs. machine; it's about using every tool at our disposal to deliver insight and impact—and ultimately, to better serve our clients while staying true to the art of consulting. The journey is just beginning, and it's one we can approach with confidence. A recurring theme in this book is that our value isn't being diminished by AI; when we play it right, it's being *amplified*. The consultants who thrive will be those who embrace this moment of renewal, pair their timeless human skills with powerful new technology, and step into the future with a mindset of adaptability and purpose. That's the adventure that awaits—and it's an exciting time to be a consultant.

ROB BERG

CHAPTER TWO

◆

The AI-Augmented Consultant Mindset

I n the age of intelligent systems, the greatest differentiator isn't fluency in AI tools—it's clarity about your authentic value, what you bring to the table that AI cannot. This chapter is about mindset—not mindset as motivational fluff, but as the deep, strategic orientation that determines how consultants will adapt, evolve, or be left behind. It's worth repeating that the consultants who thrive in this new era are not the most technical; they're the most intentional and human-centered in their approach.

Leading voices in AI echo this emphasis on human value. Meta's Chief AI Scientist Yann LeCun envisions AI ushering in a "New Renaissance" by amplifying human intelligence rather than replacing it. Similarly, Fei-Fei Li, co-director of the Stanford Institute for Human-Centered Artificial Intelligence (HAI), reminds us that the most important use of AI is to *augment* humanity, not replace it. In other words, success with AI will come from leveraging the technology to elevate our human strengths, not from trying to compete with the machines on brute force or data alone.

Beyond the Binary: Leverage, Not Replacement

Let's dismantle the false dichotomy at the heart of much AI discourse—the idea that AI will either replace consultants or leave them untouched. The truth is more nuanced—and far more empowering. Recall Andrew Ng's assertion that it's people who fail to adopt AI that will be replaced by AI; those who learn to leverage AI will outperform those who ignore it. Reid Hoffman, co-founder of LinkedIn, frames it as a matter of tooling: "If you don't use

AI you will be under-tooled and won't be competitive." The message is clear: the question isn't *whether* AI will change consulting, but *how* you will change alongside it.

AI doesn't replace consultants; it replaces unexamined workflows, outdated assumptions, and low-leverage tasks. It shines a light on inefficiencies and exposes where consultants spend time on work that doesn't require their unique expertise. This shift in work pattern reflects Li's principle of augmentation over replacement, and we see it in practice every day.

Reclaiming Strategic Time. Let's talk about Tasha, a (fictionalized) solo consultant specializing in nonprofit strategy, who used to spend five hours preparing customized discovery documents before each engagement. With AI support, she now spends forty-five minutes—and reinvests that reclaimed time in deeper stakeholder interviews. The result? Stronger proposals, faster onboarding, and more meaningful client connections. This is the AI dividend: not doing less, but doing more of what matters. Instead of cutting her workload, Tasha has amplified her impact. Consultants no longer win by hoarding information; they win by orchestrating insight. In an era where factual knowledge is cheap, true value comes from context, judgment, and the creative application of knowledge—things only a human consultant can deliver.

Human + AI: Better Together

Think of the most effective consultants you know. They ask better questions. They listen more deeply. They make clients feel seen, heard, and stretched. None of these traits are automated—or automatable. But AI can scaffold these abilities, extending your reach and enhancing your effectiveness without replacing the human touch. For example, intelligent assistants can:

- Suggest follow-up questions you may not have considered, sparking deeper inquiry.

- Summarize multi-party conversations so you can stay fully present in meetings instead of scribbling notes.

- Map out stakeholder dynamics or sentiments from communications, giving you better situational awareness before a call.

The goal isn't to outthink the machine; it's to use the machine to expand your human bandwidth. With AI handling the grunt work of information processing, you can focus more on empathy, strategy, and relationship-building.

Importantly, evidence is emerging that human + AI teams dramatically outperform humans or AI alone. In a 2023 field experiment at Boston Consulting Group, consultants with access to GPT-4 completed more tasks, more quickly, and at higher quality than those without AI assistance. Wharton professor Ethan Mollick, who studied the results, noted that consultants benefited across the board—even relative novices were able to produce work approaching expert quality when amplified by AI. The research identified different styles of effective human-AI collaboration: some consultants operated like "centaurs," delegating distinct subtasks to the AI, while others became "cyborgs," integrating AI into every step of their workflow. There's no single formula for partnership, except that the best consultants find a way to make AI a teammate.

This human + AI synergy is the real opportunity. AI can be your ever-ready junior analyst, brainstorming partner, and tireless organizer. It delivers a force multiplier, allowing a single consultant to achieve what used to require a team—not by removing the human from the equation, but by freeing the human to do what they do best. "AI is a huge intelligence amplifier," LinkedIn cofounder Reid Hoffman observes, automating the repetitive and accelerating the creative. When you embrace this mindset, you stop worrying about AI doing your job, and start leveraging AI to do your job better.

The Four Augmentations of AI-Enhanced Consulting

When used wisely, AI doesn't narrow the consultant's role; indeed, it expands it. We define four domains of augmentation that create real value. These are like the points of a compass guiding you toward higher-impact work:

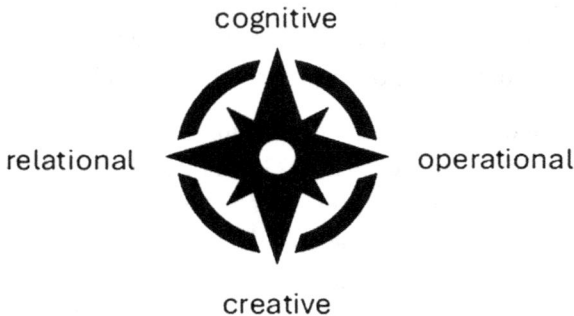

cognitive

relational　　operational

creative

Which of these four domains is least developed in your practice today?
Where could a 1 percent experiment in that area create leverage?

Cognitive Augmentation. This is where most consultants start—using AI to summarize, synthesize, and analyze information faster. With the right prompting, consultants can hand off tedious analytical tasks and focus on interpretation and decision-making. For instance, you might analyze RFPs to identify hidden client priorities, extract key insights from a 150-page report in minutes, or generate a pro/con table for strategic options at the drop of a hat. For example, a market strategy consultant could use ChatGPT to sift through several years of customer feedback to identify a previously overlooked pattern wherein clients valued speed of service even more than cost. That type of insight could reshape your messaging and help you to close more business. AI won't provide the final answer, but it *will* accelerate the path to a critical insight.

Creative Augmentation. AI can't create genius, but it can unblock it. In creative work, AI is an always-on brainstorming partner. It can generate naming ideas, analogies, or metaphors on demand. It can propose strategic scenarios or even draft the bones of a narrative, allowing you to focus on refining the best parts. It can help consultants reframe problems in novel ways by offering outside-the-box perspectives synthesized from diverse sources. For example, if you're finding ideas starting to stagnate during a client brainstorming session, you might invoke ChatGPT to suggest a few metaphorical narratives to support a proposed solution to a pressing client problem. One of those ideas might resonate so strongly that it seals the deal, selling the client on the solution. In this scenario, AI doesn't replace the consultant's creativity; it expands the palette of ideas to draw from.

Relational Augmentation. AI won't build relationships for you, but it can deepen them by equipping you with better tools and insights. Relationship management is a deeply human domain that involves trust, empathy, and credibility, yet AI can support it in subtle but powerful ways. For example, you might use AI to create stakeholder empathy maps gleaned from sentiment analysis of emails or interviews. AI can generate pre-meeting briefs highlighting each stakeholder's background and likely concerns, so you walk into the room extraordinarily prepared. It can provide real-time transcripts and summarize multi-person discussions, allowing you to catch nuances or emotions you might have missed. For example, when supporting change management you might utilize a custom GPT-powered analyzer to comb through hours of stakeholder interviews, detecting emotional undercurrents. The AI might flag that a particular department head consistently uses anxious language when discussing a certain process change, which would be a cue to address unspoken fears. Armed with these AI-derived insights, you can navigate resistance with far greater tact and empathy. The relationship still depends on human-to-human connection, but AI serves as a coach in the background, whispering helpful hints in your ear.

Operational Augmentation. This is the most visible (and often most overlooked) layer of consulting work—the administrative, procedural, and operational tasks that support your business. Here, AI and automation can save significant time and reduce friction. Think automated proposal generation based on past successful proposals, or automatically drafting follow-up emails and status updates. Scheduling, onboarding, invoicing, and CRM updates can be streamlined with AI assistants or integration tools. For example, if you find yourself spending too many hours on administrative tasks, you might set up an AI-driven workflow (e.g., using n8n and OpenAI APIs) to auto-generate pre-meeting emails, compile agenda documents, and even draft invoices. In doing so, you'll likely save many hours per week that can be reinvested in client strategy sessions and business development. The key here is to follow Andrew Ng's advice from his keynote address at ScaleUp:AI 2024: "Govern applications, not technology. Automate tasks, not jobs." In operational augmentation, we're not eliminating the human role; we're eliminating busywork. The consultant's *job* remains, but stripped of tedious tasks, you can focus on higher-value contributions.

Historical Parallels: Previous Inflection Points

We've seen transformative technology shifts before. Every major evolution in consulting has sparked fear and opportunity in equal measure. Consider a few parallels from history:

- **Spreadsheets (1980s).** When electronic spreadsheets like Lotus 1-2-3 and Excel emerged, they replaced armies of manual number-crunchers and ledger keepers. Many feared consultants or analysts would become obsolete when anyone could plug numbers into a PC. In reality, spreadsheets elevated the role of consultants who knew how to interpret models and ask the right "What if?" questions. The grunt work of calculation was automated, but demand surged for strategic analysis and modeling expertise. Consultants who mastered the new tools thrived by delivering deeper insights to clients with far greater speed and accuracy.

- **The internet (1990s–2000s).** The internet ended the era of information asymmetry. Suddenly, clients could Google anything—market data, best practices, competitor reports—on their own. This threatened the traditional consulting value proposition of being the expert with all the answers. But it birthed a new need for information synthesis and narrative development. With information superabundant, clients got overwhelmed. Consultants evolved to add value by filtering signal from noise and crafting compelling stories that made sense of complex information. Knowing *what* to do became less of the challenge; knowing *why* and *how* (and getting everyone aligned around that narrative) became the new differentiator.

- **Offshoring (2000s).** As executional tasks (research, PowerPoint building, basic analysis) could be sent offshore or to subcontractors, consultants in high-cost markets had to let go of being hands-on do-ers for every task. This shift forced consultants to more firmly take ownership of key insights, group facilitation, and organizational alignment. Leading a workshop that unearths uncomfortable truths from a client's team, or diplomatically aligning warring executives—these relational and conceptual skills became more important as the mechanical tasks were delegated away. The consulting role shifted up the value chain once again.

Now AI is yet another wave. It automates pieces of what we do, just as those past innovations did. Yes, it will change workflows and force us to redefine roles, but it doesn't spell the end of consulting any more than spreadsheets or the internet did. The consultants who thrive are those who keep redefining where they add the most value. Notably, LeCun has compared today's AI assistants to the invention of the printing press—a technology that "made everybody smarter" by massively amplifying access to knowledge. We can expect a similar leap in capability. AI will handle more routine work, effectively raising the baseline of what any individual can (and should) do. This means the bar will rise for what human excellence looks like. Just as everyone with Excel became a "quant" of sorts, everyone with AI will have some analytical and creative superpowers. Our job as consultants is to keep moving up that value chain. History tells us that those who adapt early and retool their mindset reap the biggest rewards.

Embracing Experimentation and Iteration

There is no "final form" of AI fluency. The tools are evolving and so must we. The most effective consultants treat AI the way they treat a new market or a new client problem; they test, learn, and adapt. They don't wait for the perfect use case or a corporate mandate. They dive in and tinker. In a word, they experiment—often and courageously. This willingness to play, to run small pilots, and even to fail in controlled ways is emerging as a new and critical consulting capability.

Think of AI as a creative medium in its own right. You wouldn't become a great presenter without ever stepping on stage, and you won't become an AI-augmented consultant without ever invoking an AI tool in your daily work. Wharton's Ethan Mollick, who studies and teaches about AI in business, encourages professionals to engage with AI as a "co-worker" and coach—which implies hands-on collaboration. You learn what these tools can do (and where they stumble) by actually integrating them into projects. It's iterative; you try a little automation in your proposal writing here, attempt an AI-generated slide outline there, see what works and what doesn't. Each cycle of experimentation hones your technical skill *and* your strategic instinct for where AI adds value vs. where the human touch is irreplaceable.

Small bets that pay off. To extend this idea into practice, here are a few ideas to get yourself thinking—and doing—more with AI:

- Try a new AI tool on internal work before using it live with clients. For example, use an AI scheduling assistant for your team meetings to work out kinks before deploying it with a client.

- Use AI to draft a blog post or analysis, even if you don't publish it. The goal is to practice giving effective instructions (prompts) and editing AI outputs. It's like batting practice for your prompting skills, with no pressure.

- Run a "time audit" of your week to identify low leverage work that AI could absorb. Are you spending four hours formatting slides or compiling status reports? That's your cue to find an AI or automation that can do it for you. Start with a thirty-minute task and see if AI can handle it.

The Iterative Edge

Let's say you decide to try out an AI transcription and summarization tool during client meetings. Your first attempts may feel awkward and intrusive. But if you keep at it and adjust your approach with each successive meeting, you'll eventually refine your prompts to the point where they accurately summarize meeting highlights and even reveal emotional cues from meeting transcripts. Eventually, you're likely to reach a point where you stop taking manual notes altogether. Freed from scribbling, you might find your live conversations with clients deepening as you become fully present. What might have felt like a clunky test at first could evolve into a major differentiator for your practice. The main takeaway is that while the first try with an AI tool might be the worst it will ever be, if you push through the discomfort, each iteration yields ever-higher learning. In the age of AI, it's good practice to regularly try out new AI features in low-stakes settings to keep yourself on the cutting edge of what's possible.

This experimental mindset keeps you agile. It's exactly what early adopters in our field are doing. By contrast, waiting until AI is 100 percent proven or until someone hands you a training manual means falling behind. Start using AI *now* to get those amplified benefits; otherwise you risk losing agency in your own profession. The sentiment might sound bold, but it reflects reality. Most organizations have already embraced AI, and the pace of improvement is relentless.

The Inner Work of AI Adoption

Adopting AI isn't just a technical endeavor. It's an existential imperative. Many consultants built their professional identities around being the master of their craft, the expert with the answers, the person who can crunch the numbers or craft the perfect slide deck. When a machine can mimic parts of that mastery in seconds, it can trigger a real identity crisis. We find ourselves asking, "If the AI can do X (some task I've mastered), then what is my value now?" "Who am I in this new equation?"

The answer is that our value lies in what the AI *cannot* do. It lies in judgment, integrity, creativity, and presence. AI challenges us to shift from being knowledge holders to wisdom cultivators. We move from having answers to asking better questions. From performing as the know-it-all to coaching others through complexity. We learn to hold the space in a client conversation rather than fill it with slides. In essence, we double down on the human dimensions of consulting. This is a profound mindset shift: your job is no longer to be the repository of information (because AI is a far bigger repository); your job is to be the sense-maker, the trusted advisor, the truth-teller, the empathic guide. In educational terms, you become more of a coach or mentor—drawing out insights through powerful questions—rather than a lecturer delivering facts. This approach is exactly what will set consultants apart in an AI-driven world.

It's not necessarily easy. There's Inner work to be done. You might have to let go of some ego attachments ("But I was the Excel guru! I was proud of that skill…"). You might need to unlearn the habit of equating your worth with how hard you labor or how much information you personally control. In his recent book, *Superagency*, Reid Hoffman notes that when confronted with AI's abilities, people initially feel a loss of agency. He states that "You don't want to change but you can't choose not to," in the face of AI disruption. But if you push through that phase, a new perspective emerges, where "repetitive tasks might be automated, creative processes accelerated...You get a lot more agency and so do other people." In other words, by relinquishing control over the trivial stuff, you actually gain freedom to focus on what matters most. You get to step into a higher version of your role.

So ask yourself, "If AI handled 50 percent of my workload tomorrow, how would I reinvest that time to deepen my value as a consultant?" This reflection isn't hypothetical; it's practical. Let that question guide you as you redesign your approach—not to survive AI, but to lead through it. The consultants who treat AI as an invitation to elevate their role (rather than a threat to it) are the ones cultivating true wisdom. They are practicing adaptability, empathy, and creative thinking in new ways.

Wisdom Over Wires

During a recent engagement, a client expressed his satisfaction at how I had "an uncanny ability to ask just the right question at the right time." He further said that "We could really use more of that here," confiding in me that previous consultants loved to tell them everything they knew even if they didn't want to know it. Flattered, I just smiled and said, "Thanks; I just think it's important to be a catalyst, not a bloviator."

This little anecdote carries a big message. AI can help you prepare. It can help you analyze and even help you reflect. But only you can hold space in the moment. Only a human being can truly pay attention in that profoundly present way—to read the room, sense the unspoken, adapt on the fly, and connect heart-to-heart. That's your edge. That's what clients remember long after the PowerPoint deck is forgotten. And that's what no machine, no matter how advanced, can replicate.

As we integrate AI into our consulting practice, let's keep that human core alive and well. Our tools may get immeasurably smarter over the coming years, but our mandate as consultants remains to serve our clients with integrity, creativity, and empathy. The AI-augmented consultant mindset is ultimately about amplifying humanity. It's about using every tool at our disposal to create more wisdom, more insight, and more value for the people we work with. As Stanford's Li and other thought leaders emphasize, we must put human well-being, dignity, and jobs at the center of this AI revolution. We must harness the wires (the technology) in service of wisdom (our human judgment and values).

Embracing that mindset will surely help us to survive the rise of AI *and* lead and inspire through it. Our authenticity becomes our differentiator. Our courage to experiment becomes our catalyst for growth. And our relentless focus on the human element becomes the beacon that guides clients through uncertainty. That is the promise of the AI-augmented consultant: wisdom over wires, humanity over hype. It's both a strategy and a philosophy of practice. And it's how we'll continue to elevate our profession in the years ahead, no matter what new tools come along.

No algorithm can replace the consultant who has the mindset of a lifelong learner, a creative experimenter, and a compassionate coach. That is our future. Let's dive in with confidence and purpose, using AI to become *more human*, more visionary, and more impactful than ever before.

Identity Check-In

Before we move on, take a moment to reflect on the following questions:

- Where do you currently add the most human value in your work? (Think about interactions or outcomes that happened because of your personal touch or judgment.)

- What are you clinging to that AI could help you release? (e.g., insisting on doing something manually because it's familiar, even though it's not high value.)

- How would you redesign your client experience if you suddenly had ten extra hours per week free from drudgery? (What would you spend more time on? What would you stop doing?)

By working through questions like these, you can pinpoint the mindset shifts needed to thrive alongside AI. This is about becoming more *you*, not less—leveraging technology to amplify your authenticity and wisdom.

◆

CHAPTER THREE

The AI Practice Strategy Blueprint

A I is a strategic choice—an enabler and accelerant. This chapter helps consultants to develop an intentional, practice-wide blueprint for AI adoption. Rather than chasing every new tool, we'll explore how to identify high-leverage activities, rethink service delivery, and integrate AI into the very structure of your consulting model. In doing so, we keep the focus on human-centric consulting, using AI to amplify your impact, not to replace the personal touch clients value.

This is where curiosity meets commitment, where the consultant moves from experimentation to transformation. By the end of this chapter, you should see how deliberate AI integration can reinvent your practice design, shifting AI from a novelty to a core capability aligned with your authentic consulting style.

From Curiosity to Capability

Many professionals have begun experimenting with AI in small ways, including drafting emails, summarizing meetings, researching new topics with tools like ChatGPT or other generative AI assistants. Indeed, according to a recent Globalization Partners study, *AI at Work 2025*, some 74 percent of executives view AI as critical to the success of their company, using AI for about 40 percent of their work on average. That's a great start. But experiments without strategy produce little more than sparks without proper kindling—they flare briefly and then fizzle out. Occasional use might save a few minutes, but it won't fundamentally improve your consulting practice unless it's guided by a broader plan.

Moving from ad hoc curiosity to lasting capability means formulating an AI strategy. An AI strategy turns occasional use into consistent advantage. It creates alignment between what you offer, how you deliver it, and how you grow your capacity to deliver more. For a boutique firm or solo consultant, this strategy isn't about doing the same things faster just because you can; instead, it's about intentionally redesigning parts of your work so that AI amplifies your strengths and frees you to focus on what you do best (like building relationships and providing insight). Remember, AI should enhance, not replace, the human elements of consulting. The goal is to use AI to get rid of low-value drudgery so you can spend more time on high-value, human-centric activities like client engagement, creative problem solving, and coaching conversations. You might think of AI not as a department or a separate toolset, but as an energy source that flows through your entire practice. Where in your work is energy being wasted on repetitive or low-value tasks? Where could AI deliver leverage by doing the heavy lifting, so you can redirect your energy to areas that truly matter? To illustrate the shift from curiosity to capability, consider the following scenario:

Jared, a mid-career operations consultant, was an early adopter of AI tools. He used ChatGPT to speed up research and dabbled with Notion's AI features for note taking. These experiments were helpful, but they were scattered. It wasn't until Jared took a step back and mapped his entire client journey from initial discovery to final delivery that he realized a significant chunk of his work hours were consumed by repetitive, low-value tasks. This realization was a turning point. In just a few months, he went from playing with AI to applying it systematically. He built a set of GPT-powered standard operating procedures (SOPs) for common tasks, streamlined client onboarding, and even reduced proposal turnaround time by half. What began as casual curiosity became a capability advantage for his practice. Jared freed up hours each week, allowing him to invest more time in strategic planning and client relationships. The result was increased efficiency with a noticeable improvement in client satisfaction and business growth, all because he had a strategy for where and how to use AI.

This little vignette illustrates an important point. When curiosity is guided by a clear strategy, it leads to transformation. Many consultants are lighting little AI "sparks" in their workflow; it's the strategy that turns those sparks into a sustained fire of progress. In the next sections, we'll construct your AI strategy blueprint step by step, starting with identifying where AI can help the most.

Identifying High-Leverage Activities

To design your AI strategy, first review your current workflow. Not every process should be automated, and not every shiny new AI tool adds value. Begin with a diagnostic mindset. Ask yourself where AI will give you the best bang for your buck. In smaller consulting practices, our scarcest resource is time. So we need to deploy AI in ways that multiply that resource. This means identifying the high-leverage activities—the tasks where AI can make a significant difference (and equally, recognizing tasks where AI should *not* be applied). One way to do this is to classify your work into three core areas:

- **Repetitive tasks** are recurring tasks with predictable formats or triggers (e.g., scheduling meetings, sending onboarding emails, compiling status reports, etc.). They don't require deep expertise every time, yet they eat up portions of your day. For example, routine scheduling and follow-up emails are ripe for automation. If you find yourself doing something repeatedly in a similar way, that's a strong candidate for AI or at least some form of templated assistance.

- **Time-intensive tasks** include high-effort work with relatively marginal value-add for the client. These could be tasks like consolidating data from multiple sources, formatting documents, transcribing meeting notes, or doing first-pass research. They have to get done, but they don't necessarily require your highest consulting skills. Often, these are the behind-the-scenes duties clients never see, but which cost you hours. AI tools can dramatically compress the time needed for many of these tasks.

- **Strategic tasks** include the work that *does* require judgment, expertise, relationship-building, or emotional intelligence (e.g., facilitating a client workshop, synthesizing insights into a recommendation, coaching a client through a tough decision). These are usually high-value activities that define your consulting impact. This is where AI *supports* or augments you, and is not a replacement for your hard-earned expertise. You might use AI to gather background information or generate options, but the last mile—applying context, wisdom, and human touch—remains with you. These tasks often differentiate your services and build trust with clients.

Once you've categorized your activities, you'll have a clearer picture of your practice. You might discover, as Jared did, that a large chunk of your week is spent on things that don't fully require your expertise. Conversely, you'll pinpoint the moments that do require your full presence and should be protected from over-automation. The next step is to decide, for each category, how to treat it in your AI strategy. For that, we introduce a simple framework.

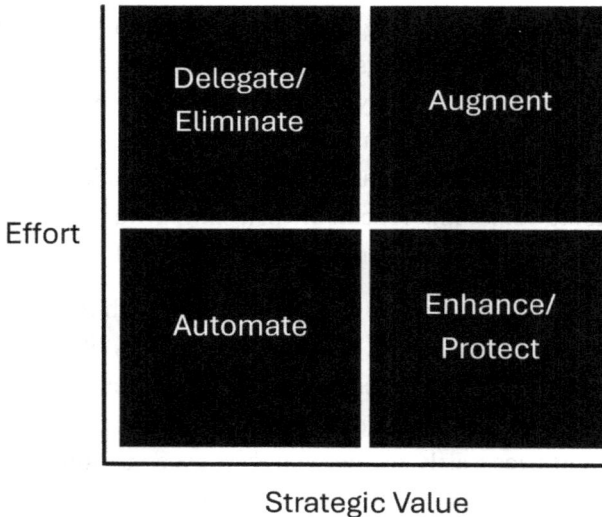

Delegate/ Eliminate	Augment
Automate	Enhance/ Protect

Effort / Strategic Value

The AI Leverage Matrix

The AI Leverage Matrix

Visualizing where to apply AI. Imagine plotting your various tasks on a two-by-two grid. One axis represents the strategic value of the task (low to high); the other represents the effort required (low to high, in terms of time or labor). This creates four quadrants, each suggesting a different approach to AI integration:

- **Quadrant 1: Low strategic value, high effort.** These tasks are high-effort busywork that doesn't greatly benefit the client or your differentiation. For items in this quadrant, the best strategy is to **eliminate or delegate** them if possible. If a task doesn't add much value and is taking significant time, consider whether it needs to be done at all, or delegate it (to a junior team member, a virtual assistant, or an automated process). For example, a detailed data reconciliation that no client ever sees could be cut down or outsourced. If it must be done, this is prime territory for heavy AI automation or offloading entirely.

- **Quadrant 2: High strategic value, high effort.** These tasks are important but time-consuming. Here, look to **augment** with AI or do preparatory work manually to reduce the burden. The goal is to maintain quality and insight (since these are high-value activities) while using AI to handle sub-tasks or provide a draft. For example, preparing a strategy presentation is high-value and requires your insight, but you could use AI to generate initial slides or research data points, then you add the expert commentary. AI becomes your analyst or assistant, speeding up the work but not taking it over. You leverage AI to increase your throughput without compromising on judgment.

- **Quadrant 3: Low strategic value, low effort.** These tasks are minor chores that don't take much time individually, but could be automated to free up mental space. The mantra here is **automate** fully. If something isn't particularly valuable and is easy to automate, set up an AI or software to handle it end-to-end and don't spend your precious minutes on it at all. For example, formatting a document or

31

sorting routine emails falls into the category. Modern AI and software tools can often handle these with minimal setup. This quadrant is about quick wins—small automations that collectively add up to significant time saved.

- **Quadrant 4: High strategic value, low effort.** These are tasks that are crucial to your value proposition but don't take a ton of time—often things you're naturally efficient at or enjoy doing yourself. Here, the strategy is to **enhance or protect** these tasks for human delivery. In other words, you likely want to keep doing these personally (because they define your consulting expertise), and only use AI in a minimal way to enhance the output if at all. For example, a monthly executive briefing might only take you an hour to prepare because you know the client well; the value is high because it strengthens the relationship. You might use AI to check facts or polish language, but you'd *protect* the core of this task as a human-to-human interaction. The key is not to let automation dilute the personal touch or insight that makes this work high-impact.

This mental model helps decide what role (if any) AI should play for each type of work you do. To make this concrete, let's apply the matrix to a couple of typical consulting activities:

- **Competitive research.** Often, compiling competitive or market research is *low strategic value* (it's background info, not your unique insight) and *high effort* (could take many hours). According to our matrix, this falls in the *Delegate/Eliminate* zone. Indeed, many consultants now use AI tools to handle initial research. For instance, you might use ChatGPT to draft in-depth research briefs for client reports, generating structured outlines and preliminary insights so you can focus more on refining strategic recommendations. This is a great example of delegating a heavy lift, where AI is used to scrape the info and present a starting analysis, and you review and apply the human judgment on top.

- **Strategy presentation refinement.** Crafting the story and recommendations in a strategy deck is *high strategic value* (it's core to your advisory role), and it can be *high effort* too. If it's taking dozens of hours, you're in the "augment" with AI territory. You might use an AI slide generator like Gamma, or an AI writing assistant like ChatGPT Canvas to draft text for each slide, based on your outline. You remain the director by editing, rearranging, and injecting expertise, but AI serves as your assistant, handling the grunt work of creating draft slides or checking grammar. The result is you maintain quality while significantly reducing time spent.

Into which quadrants do your various activities fall? Take a moment to reflect and sketch out a matrix for your own tasks. Here's how: For one week, track how you spend your time each day. Jot down the tasks and roughly how many hours go into each. Then categorize those tasks by the lenses above (repetitive, time-intensive, strategic) and plot them conceptually on the effort vs. value matrix. Once you see them laid out, tag each activity with one of four labels (eliminate/delegate, automate, augment, or enhance/protect). This simple audit can highlight immediate opportunities to reclaim time. Perhaps you'll find you spend five hours on something that could be done in thirty minutes with an AI script, or you'll reaffirm that certain high-touch tasks should never be automated. Use these insights as the basis for your AI adoption plan.

By identifying high-leverage activities and categorizing them, you set the stage for a strategic approach. In the next section, we'll look at how to redesign your workflows accordingly, deciding what to delegate, automate, augment, or leave alone in the day-to-day operation of your consulting practice.

What to Delegate, Automate, Augment, and Leave Alone

An effective AI strategy isn't about replacing entire processes wholesale; it's about redefining workflows by weaving automation and intelligence into specific steps. Think of your consulting projects as sequences of steps from first contact with a client to final deliverables. Now, consider how each step might be reimagined with AI in the mix. The idea is to build a new workflow where mundane parts flow through AI, while you step in at the points that require human judgment, creativity, or empathy.

Let's walk through a sample client engagement workflow and see where delegation, automation, or augmentation makes sense vs. where the human touch must remain. Imagine a typical engagement timeline with these major steps: lead qualification, discovery analysis, client workshops, proposal generation, executive briefing, and post-project debrief/synthesis. Here's how we might allocate our approach:

- **Lead qualification.** Early in the funnel, you often have many incoming inquiries or leads to sift through. This is largely a repetitive data-gathering and filtering task, ripe for automation. You can automate initial lead qualification by using AI-powered chatbots or screening questionnaires that capture key information, or even a simple AI that ranks leads based on fit criteria. For example, some independent consultants use AI-driven forms or email filters to triage inquiries, so they only spend personal time on the most promising leads. This not only saves time but can also respond to prospects faster (an AI assistant can reply instantly with a Calendly link or a request for more info, even while you sleep).

- **Discovery analysis.** The discovery phase (analyzing client-provided data, context, needs) is often where you exercise your judgment and experience. While you might delegate to AI to help gather background data or summarize documents, the core analysis—understanding the client's unique situation and identifying nuanced insights—is a high strategic value activity best led by you. You might augment your work by using AI tools to transcribe and summarize stakeholder interviews or to pull industry benchmarks, but the interpretation of that information and reading between the lines is a human task. Insight requires empathy and context, which AI lacks. So, for discovery, rely on your own analysis, using AI only in supportive roles (e.g., identifying sources and generating a summary of a large Excel file so you don't miss a pattern).

- **Client workshops.** Live workshops or meetings with clients are replete with real-time interaction, which is precisely where human skills shine. During a workshop, your role is facilitator and coach, responding to group dynamics, emotions, and cues in the room (or on Zoom). Of course, you would never automate the facilitation of a workshop lest you wish to alienate clients and defeat the purpose.

However, AI can quietly augment in the background. For instance, you could have a transcription tool (like Fathom, Otter.ai, or Fireflies) record and highlight key points in real time, or a GPT-based tool that suggests next questions on your screen based on the conversation (almost like a copilot feeding you info so you can stay focused on people in the room). After the session, AI could help by summarizing the workshop notes and extracting action items. In short, the delivery is human, but AI acts as an assistant *before* (prepping materials or agenda from past data) and *after* (documentation and analysis). This approach ensures the clients feel seen and heard (a machine isn't running the meeting), while you still benefit from AI efficiency behind the scenes.

- **Proposal generation.** Crafting proposals is a prime example of work that can be sped up with AI without hurting quality. Much of a proposal consists of standard sections (your approach, bios, past results) that you might be rewriting each time. You might use AI to augment your workflow by generating a first draft by pulling from your knowledge base of past proposals and tailoring it to the new client's context. For example, you might maintain a library of proposal prompts where you input the client's industry, desired outcomes, and key challenges, and an AI tool produces a well-structured draft proposal or at least an outline. This can easily cut proposal writing time by 50 to 80 percent. You, of course, will review and refine the draft to ensure it aligns with your voice and the client's specific nuance, which is where the human finesse comes in. But think of the AI as a junior proposal writer who never gets tired and is never offended by your ongoing critiques to get it just right. I've personally automated at least 60 percent of my proposal and statement-of-work drafting using templates and ChatGPT, drastically reducing the delay between the promise and delivery of a formal proposal to an eager prospect.

- **Executive briefings.** Executive briefings or readouts are high-stakes moments where you're conveying the crux of your advice to top decision-makers. The relationship and trust you build in these briefings are as important as the content. As such, the consultant should remain front and center. However, AI can be used to augment the preparation. For example, you could use an AI tool to condense a fifty-page report into a one-page executive summary, then refine that summary into talking points in your own style. You might also use AI to generate data visualizations quickly from raw data to include in your presentation. But when it comes time to deliver the briefing, you're the one in the room (or on the call), reading the room and adjusting on the fly. AI's job here is to ensure you walk into that meeting with high-quality materials and insights at your fingertips. In some cases, consultants even use an AI in the background during Q&A, perhaps quickly querying a chatbot trained on the project data if an unexpected question comes up, providing a suggested answer or reference on the spot. The client, however, still sees *you* as the expert speaking directly to them.

- **Debrief and synthesis.** After finishing a project, it's a best practice to conduct a debrief where you synthesize what went well, what could be improved, and capture lessons learned. This reflective process provides high strategic value for improving your practice and providing valuable insights to the client. While such debriefs tend to be qualitative and require those uniquely human qualities of honesty and humility, you might use AI to help gather feedback (for example, analyzing a client feedback survey and highlighting common themes) or help organize your presentation, but interpreting that feedback and deciding on changes is your job. If you write a post-project report or blog about the engagement, AI can help polish your writing or organize your thoughts, but the substance comes from you. In essence, any task involving learning, sense-making, and building deeper relationships (like a candid debrief with a client) should remain largely human-driven. Use AI as a thought partner. Consider asking ChatGPT to suggest possible improvements based on the project log. It might surface an idea you hadn't thought of; but you make the call on what to implement.

This workflow breakdown demonstrates a balanced approach, where we delegate the grunt work wherever feasible, automate the rote and repetitive, augment the complex, and preserve the essential human elements where appropriate. As you consider each step in your consulting process, ask yourself a few key questions:

- **Is this task primarily repetitive or does it require judgment and nuance?** (This helps determine if it's a candidate for automation or not.)

- **Does this activity require real-time interaction and empathy (synchronous human presence), or can it be done asynchronously?** (If it's asynchronous, AI might handle it offline; if it's a live interaction, AI's role should be minimal and supportive.)

- **Will using AI here amplify my impact or could it erode the personal touch?** (Be brutally honest; if automating something might make the client feel less cared for, think twice.)

When to Say No to AI

Not all efficiency is helpful. A cautionary tale comes from a consultant who decided to automate weekly client check-in emails. She set up an AI system to automatically send personalized-sounding messages to her clients inquiring about their progress or any issues they may be encountering, thinking it would save time and promote greater client engagement. And while it did save a few minutes each week, she soon noticed that client engagement had actually *dropped*. Clients were replying less to her check-ins and seemed less enthusiastic in real-time meetings. Confused, she reached out to a couple of trusted long-term clients, who told her the automated emails felt a bit hollow. They missed the personal nuances in her usual check-ins—those little references to last week's conversation or a mutual joke. The AI version, while grammatically perfect, lacked that genuine human touch, and clients could tell. The lesson? Just because you *can* automate something doesn't always mean you *should*. In consulting (especially for solo practitioners or small firms where relationships are your lifeblood), AI must support the human touch, not replace it. If a particular automation makes a client feel less seen or valued, that's a red line. In those cases, it's better to keep doing it the "inefficient" way yourself or find a different approach altogether. Efficiency means nothing if you lose client trust in the process. Use this story as a reminder to always balance tech with empathy; the efficiencies gained from AI should be reinvested into deeper human connection, not used to distance yourself from clients.

Having mapped out what to delegate, automate, augment, or leave alone, the next step is to incorporate AI into your actual service design—the offerings and value you deliver to clients. We'll explore how AI can enhance the products and services you provide, making your practice not only more efficient but also more distinctive to your clients.

Building AI into Service Design

Not only can AI make your internal workflow smoother, it can also redefine how you show up to clients and the value you offer them. The idea here is to embed AI into your service offerings in a way that enhances the client experience and outcomes. This is a shift from using AI privately as your secret productivity tool, to using AI *openly* as part of your value proposition. For small firms and solo practitioners, this can be a real differentiator. It can make a small practice feel larger and more capable than it is, all while staying customized and personal.

To spark your creativity, here are a few examples of how consultants are weaving AI directly into their client-facing services:

- **Proposal companion.** Imagine every proposal you send to a prospect comes with a little something extra, such as an AI-generated research brief tailored to that client's industry or problem. For example, if you're proposing a marketing strategy project to a mid-size tech company, along with your standard proposal document, you include a one-page AI-generated industry insight report. It might highlight recent market trends, competitor moves, or relevant case studies, all curated for that prospect. This "proposal companion" AI brief shows the client you've done your homework and provides immediate value before you've even been hired. It's like offering a free preview of insights your client can expect by working with you. The AI does the heavy lifting of generating the customized research (drawing from up-to-date sources that you specify), and you review it to ensure accuracy. Not only does this save you hours of research time, it impresses the client. It sets you apart from larger firms that might not bother to personalize at this stage. You're demonstrating curiosity and effort upfront. Such touches, enhanced by AI, can increase your proposal win rate by making them more engaging and insightful.

- **Client session copilot.** When you're in the middle of a complex project, keeping track of every detail from past meetings and communications becomes challenging, especially as a solo or small firm consultant juggling multiple clients. A "session copilot" is essentially an AI assistant that lives in your client sessions (meetings, calls, workshops) to help you recall information exactly when you need it. Practically, this could be a laptop or tablet interface you have on the side during a meeting. It's connected to a GPT-based system that has been fed all the transcripts, notes, and decisions from prior sessions with that client (and perhaps relevant project documents). With one click it can list decisions about specific topics in your last meeting, or provide a summary of issues identified in the discovery phase of your project. This way, you can confidently facilitate the session knowing that if someone asks a question like, "What did we conclude on that budget issue two weeks ago?" you can get the answer in seconds from your AI copilot. It's like having an encyclopedic memory on hand. This not only makes you look extremely prepared and attentive, it frees your mental bandwidth to focus on the human dynamics in the room, since you're less worried about remembering every detail. Some consultants even set up triggers. For example, say "Hey Jarvis" (if your AI is nicknamed Jarvis) and ask a question aloud to retrieve info, almost like having an analyst in the room. The key is that the client sees *you* seamlessly recalling facts and building on them—they don't even need to know an AI is feeding you background info. Or you might choose to be transparent and say, "Let me just consult our session summary," and show that you have this advanced tool, which can itself build confidence that you're tech-savvy and thorough.

- **Follow-up GPT.** After a project is delivered, clients often have follow-up questions. They might be reading your report and want clarification on a particular data point, or a month later they wonder how a recommendation should be implemented in a new situation. In a traditional setting, they'd email or call you, and you'd respond when you can. But what if you could offer them a 24/7 on-demand advisor in the form of a custom AI chatbot trained on the specifics of their project? This is what a "Follow-Up GPT" entails—essentially a branded AI assistant you provide to the client for some period of time after the engagement. For example, you finish a consulting project and deliver a final report. Along with it, you deliver access to a chatbot (which could be as simple as a secure chat interface on a website) that has been trained on that report and all its supporting research. The client's team can ask this chatbot questions like, *"What was the rationale behind the third recommendation?"* or *"Can you show the data from the market analysis section?"* The chatbot will generate answers drawn only from the project data, often with citations (page references from the report, for instance). This provides extra value to the client; they have an "always-available analyst" at their fingertips without constantly tapping your shoulder. It also reduces your own load of answering routine follow-up queries. You might limit it (e.g., the chatbot is available for one month post-project as part of your service), or keep it as a paid ongoing support product. From the client's perspective, they feel they have gotten a little more than just a static report—they have an interactive tool to continue extracting value. Importantly, such a chatbot should be clearly based on *their* project data (it's not a generic chatbot giving generic advice; it's tailored to them), and you'd position it as a value-add tool. This kind of AI enhancement to your deliverables can make a one-person consultancy appear far more leveraged and responsive. It strengthens the relationship by showing you care about their success even after the formal engagement.

Each of these examples shows AI woven into the fabric of your service, not just used in the back office. They demonstrate a mindset shift; you're designing your offerings with AI in mind from the start. This can lead to innovative service models that were not feasible for small firms before. You're essentially using AI to scale yourself in a way that maintains authenticity and quality. To show how powerful this can be, here's a quick (ok, somewhat fabricated, but based on fact) story.

Lucia, a boutique strategy consultant, decided to implement a bold idea: with every major client deliverable (like the final strategy report), she packaged a custom-trained GPT chatbot that her clients could query. She trained each chatbot on the specific engagement's documents and insights. After delivering the final report to the client's executive team, Lucia would introduce their "personal strategy assistant," which was simply an AI chatbot capable of answering any question about the project's findings, complete with references to the report sections. Clients were intrigued. Over the next few months, they *loved* having this resource. The CEO could ask the bot at 10:00 p.m. on a Sunday night, *"What were the key factors in our SWOT analysis?"* and get an instant answer. The head of marketing could query, *"What did we conclude about competitor X's strategy?"* without summoning Lucia. Clients raved about this "advisor who never sleeps." From Lucia's perspective, it reduced her post-engagement email load dramatically. Many minor questions never reached her because the chatbot handled them. More importantly, it made her tiny firm look much larger. Providing an AI assistant made it seem like she had a whole team of analysts at the ready, when in reality it was just one well-configured AI service. The clients' trust in her firm grew, and many recommended her to others for the innovative way she delivered value in addition to her strategic acumen. Lucia's story shows how embracing AI in service delivery can amplify a small practice's impact while keeping the consultant's human expertise at the core. (After all, the chatbot's usefulness ultimately came from Lucia's great work and insights that it was trained on. It was her knowledge, just made more accessible.)

The takeaway for your AI strategy blueprint is to ask how AI might enable you to offer something new or extra to your clients. Think about your own services. Is there a part of your value delivery you could clone and make available on demand or at scale with an AI solution?

If you could clone one part of your value delivery and make it available any time of day or night, what would it be? That's a prime candidate for your first AI-enabled product or service extension. For example, if clients always value your quick email responses with helpful advice, an AI that provides FAQ answers might be your cloned offering. This can be extraordinarily powerful.

By infusing AI into your service design, you're not only improving efficiency, but also deepening client relationships. Clients will sense that you are proactive and forward-thinking, always looking for ways to help them more. Just ensure that whatever AI-driven component you introduce is clearly aligned with your authentic consulting style and adds real value (never gimmickry). If you pride yourself on white-glove, personalized service, your AI additions should enhance that feeling, not detract from it. Done thoughtfully, AI can help you deliver more value than ever without compromising the personal touch that defines a boutique practice.

Now that we've covered a few ways to integrate AI with standard workflows and service design, let's discuss the enabling conditions—the infrastructure and systems that you might need to have in place to make all of this work smoothly.

Infrastructure and Integration

Behind every effective AI-enhanced consulting practice is an often unseen but crucial foundation: a well-integrated, right-sized tech stack and supportive infrastructure. As a solo consultant (or even a small consulting firm), you don't need an enterprise-class IT department, but you do need to thoughtfully choose and connect the tools that will power your AI strategy. This foundation covers everything from the AI tools themselves to data management and security. Let's break down the key components you should consider when building your AI-ready infrastructure:

- **Tool stack.** This is the collection of AI and automation tools you'll use day-to-day. It might include a service like ChatGPT, Claude, or Gemini for text generation, specialized tools like Notion AI or Airtable for productivity, transcription tools like Fathom or Otter.ai for meetings, and automation platforms like Zapier, n8n, or Make to connect processes together. The key is to select tools that align with your needs and your comfort level. There are hundreds of AI tools out there, but you likely need only a handful that cover your use cases (content generation, data analysis, scheduling, etc.). Don't fall into the trap of tool overload; instead choose tools that fit your authentic consulting style, rather than what's trendy. For example, if you rarely create slide decks, you don't need an AI slide generator in your stack. If writing is your forte and part of your brand, you might use AI to support but not to ghostwrite for you. Pick a primary AI assistant (like ChatGPT) as your go-to and supplement with a few niche tools. Get them talking to each other via integrations.

- **Data governance.** With great power (of AI) comes great responsibility regarding data. As a consultant, you're likely to handle sensitive client information. Introducing AI means you must manage data carefully, both to protect client confidentiality and to abide by ethical and regulatory guidelines. Data governance includes secure storage (where are you keeping transcripts, client documents, AI outputs? Is it a secure cloud drive with proper access control?), ethical handling (are you inputting sensitive client data into third-party AI tools? If so, have you checked their privacy policies? Additionally, be careful to avoid feeding confidential details into public AI services without client permission), and transparency with clients (do your engagement terms mention you use AI assistance? In many cases it's wise to let clients know, focusing on the benefits, and ensuring them that no sensitive info will be mishandled). A striking fact from my own experience is that few companies that use AI have established policies for employees' use of generative AI while large majorities of those employees use the technology unsupervised. This indicates many are winging it, which is not a good idea when trust is at stake. For your practice, establish your own data rules. For example, "I will not upload client-identifiable data to any AI service without anonymizing it," or "I use self-hosted or enterprise versions of tools for any proprietary data." This way, you protect your clients and yourself. Good governance might not be flashy, but it's part of the professionalism that will set you apart in the age of AI.

- **Knowledge management.** As you start producing AI-assisted work, you'll find yourself creating a lot of digital artifacts, including prompt templates, useful outputs, insights generated in seconds that might have taken you hours to think of manually, etc. Don't let these valuable assets go unleveraged. Knowledge management is about capturing, storing, and retrieving the intellectual capital produced in practice. This includes both the final deliverables *and* all the intermediate work products. For instance, you might have a prompt that reliably produces a great SWOT analysis outline; save it! Or an AI-generated draft that you only used a part of, but contained other ideas that you can archive for future reference. Consider setting up a simple system, such as a folder in your OneNote/Notion or a wiki where you keep "AI Recipes" (e.g., a prompt with an example outcome) and "AI Outputs Library." Over time, this becomes your internal R&D engine, accelerating future projects. Also, ensure your system allows quick retrieval; tag things by topic or client. The benefit is twofold: speed (you're not reinventing the wheel each time) and learning (you'll learn from both successes and failed experiments). Sadly, knowledge workers often spend a great deal of time needlessly searching for information they've already seen. A good knowledge management practice, enhanced by AI (imagine an AI that can search your personal knowledge base quickly), can significantly reduce that waste.

- **Permissions and roles.** If you work with a small team or even just a virtual assistant, think about who can access which tools and data. Set up the roles and permissions in your systems. For example, if you use a client relationship management (CRM) tool that integrates with an AI, maybe only you should have access to certain client notes, while an assistant can update contact info. Or if you have an AI knowledge hub, perhaps subcontractors or junior consultants can contribute but only you curate the final knowledge entries. Permissions also matter in automation. For example, an automated email system should perhaps only send drafts to you for approval rather than auto-sending to clients until it's proven reliable. It's easier to set these boundaries early than to fix a breach later. In a solo context,

"permissions" might sound irrelevant, but even then, consider separating accounts or profiles for different purposes (a personal OpenAI
account for general experiments vs. a company account for client-
related work, for instance). This way, if you scale up in the future,
you already have a structure in place.

A final word on infrastructure is to start small. You don't need to invest in
an elaborate tech stack on day one. In fact, many successful solo consultants
run a very lean setup. (In building my last practice, I only added a full-time
employee after I hit seven figures in consulting contracts. It was subcontractors that helped me get there.) The trick is to ensure the tools you do choose
can talk to each other (e.g., through APIs or workflow orchestration tools)
so that you retain the option to automate data flow between them at some
later date. Also, be mindful of cost. Many AI tools have subscription fees or
use tokens from your LLM provider (which, depending on the LLM used,
can add up quickly), and it's easy to sign up for many and not fully use them.
Pick one from each category that suits your needs and only expand when
you find a clear gap.

Choose one core AI platform (your primary "assistant"), one reliable storage
solution for documents (e.g., Google Drive, Notion, etc.), and one workflow
automation tool (if needed). Use them until you *feel* constraints or pain
points. Only then consider adding something new. This prevents the common "tool fatigue" and ensures you get the maximum value from minimal
complexity.

Designing Your Internal R&D Engine

As alluded to above, don't make the mistake of capturing only final outputs
(presentations, reports) in your archives. By the time something is a polished
report, a lot of the generative magic and learning has already happened. In
the age of AI, develop a knowledge hub that captures that large proportion
of reusable value that lives in intermediate work, such as:

- **Prompt templates and scripts.** Keep a catalog of the prompts that
  worked well for different tasks (and notes on those that failed and
  why). Over time, this becomes a playbook you can refine. For example, you might develop a go-to prompt for analyzing survey

results, converting proposals to legal contracts, or drafting a project kickoff agenda.

- **AI-assisted email drafts and copy.** If you have AI help write marketing emails or routine client communications, save the early drafts (especially if you edited them); they can serve as templates for future work or even training data if you develop a custom model.

- **Lessons from failed experiments.** Not every AI attempt will work. Perhaps you tried to automate slide creation and it flopped. Document what went wrong in a few notes. These lessons learned are incredibly valuable; they prevent future you (or colleagues) from repeating the same mistakes and spark ideas for different approaches.

- **Micro-insights and reframes.** These are those little insights or alternative angles that came up during analysis but didn't make it into the final deliverable. Maybe an AI analysis of data revealed a curious pattern that the client ultimately didn't care about, or you brainstormed five ideas and only used three. Store the other two somewhere; they might be useful for a different project or as inspiration in the future.

By storing intermediate work products in a structured way (with tags, keywords, or in a database), you essentially create a personal "consulting brain" backed up in the cloud. Later, you could even use AI to query your own knowledge hub (e.g., prompting *"AI, retrieve all past proposal intro paragraphs for healthcare clients"* delivers instant material to work with). This kind of hub makes your practice more resilient and scalable. It's your own AI-ready knowledge base that grows over time, making each future project easier. It also aligns with the philosophy of continuous learning and humility by demonstrating that you're always accumulating knowledge in addition to producing deliverables. Over months and years, this compounding repository becomes a competitive advantage in itself.

At this point, we've covered strategy, workflows, service design, and the supporting infrastructure. But even the best strategy on paper will falter without one key element, people and mindset. In a small firm, "people" might just be you and a partner or a virtual assistant; in a boutique firm, it could be a dozen consultants. In any case, how you (and your team) embrace or resist these changes will determine success. That's why the next focus is on training and adoption to get buy-in and build an AI-embracing culture in your practice.

Training and Adoption Strategy

You don't need everyone on your team (if you have a team) to be an AI expert, but you *do* need everyone on board with the shift. If you're solo, this is about setting aside time to train yourself and forming the right habits so that AI integration sticks. If you lead a small firm, it's about change management—encouraging colleagues or employees to use the new tools and new processes you're implementing. Culture is paramount; a supportive, curious culture will eat any shiny AI tool for breakfast (to paraphrase Drucker's famous "culture eats strategy for breakfast" quote). In other words, success depends more on people's mindset than on the technology itself. To that end, here are some thoughts on driving AI adoption:

- **Identify a few AI champions.** If you have a team, find those one or two colleagues who are naturally excited about AI or at least open-minded and curious. Tap them to lead by example. They can experiment and share successes and failures candidly. In a solo practice, you can coordinate with a peer in your network; for instance, partner with another consultant to trade AI tips weekly. Being an "AI champion" means being the person who tries new things and spreads the knowhow. It's easier for others to follow if someone they trust goes first.

- **Run monthly show-and-tells.** Make it a habit (in team meetings or personal review time) to reflect on how AI contributed recently. For a team, have a monthly, fifteen-minute meeting for new ways you're using AI to save time or improve quality. This creates positive peer pressure where everyone wants to have something to share. For yourself, keep a small journal of AI wins and issues. By reviewing it monthly, you reinforce what you've learned and stay motivated by the progress. These show-and-tells reinforce the idea that AI skills are now part of professional development. It also normalizes the learning curve (when someone shares a mistake they made and how they fixed it, others realize it's okay to not get it perfect the first time).

- **Track and celebrate small wins.** Especially early on, celebrate any time saved or improvement made thanks to AI. Did the new scheduling bot book ten meetings for you with no manual emails? Awesome! That's an hour saved, which might have been spent reading a thought-provoking article or having an extra chat with a client. Did a junior team member use ChatGPT to outline a report and consequently deliver a draft a day early? Highlight that. By recognizing these wins, you tie AI usage to tangible benefits, which increases buy-in. It sounds simple, but a little acknowledgement goes a long way in habit formation. Consider keeping a scoreboard that highlights hours saved or projects accelerated by utilizing AI to make progress visible.

- **Create a safe space for experimentation.** This is crucial. Adopting AI is a learning process and there will be missteps. Make sure you and your team feel comfortable sharing when something doesn't work. Maybe an automated report had errors, or someone embarrassed themselves by trusting an AI-generated fact that was wrong. Instead of scolding or feeling discouraged, treat these as learning moments. For your team, let your colleagues know that it's okay if something goes wrong as you learn something valuable. If you're solo, be kind to yourself; you might try an AI tool that accidentally sends a weird message or messes up a calculation. Rather than conclude that AI is too risky, dissect what happened and adjust accordingly. The firms and individuals that get the most out of new tech are those that have license to play around a bit *and* calmly reflect on failures. It's all part of the process of learning and growth.

There are, unfortunately, countless stories of failed tech initiatives, and AI offers no exceptions. In one of my client companies, leadership rolled out some AI tools that were expected to boost productivity significantly. But they forgot that humans were involved, and for humans to embrace anything new, they need to buy in. Staff were still expected to produce as usual, and there was no incentive or recognition for using the tools. In fact, some felt that if they used AI to work faster, it might reduce their value as they'd be creating too much free time! The result? Minimal adoption. The tools gathered dust because no one had a real reason to change their habits. Contrast this with another company of similar size that made AI experimentation part of their performance reviews—not in a heavy-handed way, but by adding the question, "What did you try with AI this week?" in weekly check-ins. They also allotted a small training budget and gave shoutouts to people who found creative uses for the new tools. Within a few months, AI adoption soared. People were sharing tips actively, and the tools started paying for themselves through efficiency gains. The moral is clear: even if you're just one person, align your "soft" incentives with the change. If you bill hourly (a dying practice in the age of AI), think about offering some value-based or flat-fee services so that time saved by AI actually benefits you and the client, rather than making you feel like you lost income. If you have a team, consider how you can encourage them through publicly recognizing successful AI initiatives, or simply incorporate AI skill development in your professional development plans. The best tools on the market won't help if mindset and culture aren't aligned to use them.

By focusing on these adoption strategies, you create an environment where AI is not a flavor-of-the-month gadget, but an evolving part of your practice's DNA. People remain at the center—being creative, sharing, learning— exactly as it should be in a human-centric consulting model. AI then becomes a catalyst for that creativity and learning, rather than a threat or a forced mandate.

We've now constructed a thorough blueprint from high-level strategy to specific workflow tweaks, from service innovation to infrastructure, and finally the human element of adoption. It's a lot to take in, so let's summarize the key points and provide some thought-provoking questions to get you started on your own AI strategy blueprint.

The Strategy Summary

An AI strategy for a consulting practice is about far more than efficiency or playing with novel tech. Indeed, it's really about the intentional design of your practice. It's the bridge between knowing about cool AI tools and actually transforming the way you work and deliver value. Let's boil down the chapter's insights into a few fundamental principles:

- **Clarify what makes your consulting valuable.** Always start with your value proposition. What do your clients truly pay you for? Is it your unique insights, your facilitation skill, your trusted advice? Whatever it is, make sure your AI use reinforces that core value. If your value is in customized, high-touch service, use AI in ways that give you more time with clients (not to avoid clients). If your value is analytical rigor, use AI to crunch data even deeper. Know your unique selling proposition, and guard it; AI should support, not replace, the things that define your brilliance.

- **Identify where time is wasted and where attention is scarce.** Do an audit of your operations to find the friction points. Maybe you realize you spend twelve hours a week preparing slide decks (attention that could be spent thinking of recommendations instead). Or you notice you're often rushing through developing a workshop agenda because time was lost doing something automatable. These waste areas and bottlenecks are golden opportunities for AI. By pinpointing them, you know exactly where to apply automation or augmentation to liberate your time and attention. Freeing up even a couple of hours a week from low-value tasks can be transformative if you reinvest that time wisely (for example, spending those hours brainstorming new ideas for a client or simply resting to avoid burnout).

- **Align AI tools to support (not replace) your human touch.** This cannot be emphasized enough! Use AI to augment your strengths and extend your capabilities, not to do the entire job for you or to mimic you in a way that's inauthentic. If you're a great storyteller, use AI to find data points that make your story stronger, not to write

50

a generic story. If you pride yourself on personal relationships, consider using AI to remind you of client birthdays and preferences rather than sending them an AI-written note with no personality. Every tool you implement should be used in the service of making you more effective, not making you obsolete. Keep asking yourself if the tool makes your service to the client better and makes you better at what you do. If the answer is yes, it's aligned. If the answer is unclear (or if it feels like it might distance you from your client), think twice.

- **Shift from experimentation to transformation.** It's fine to play with AI out of curiosity; in fact, that's how we all learn initially. But don't stop there. The firms and individuals who will thrive are those who move beyond the novelty phase. They will integrate AI deeply into how they operate and even what they offer, thus transforming their practice. This shift means having processes and strategies, not just one-off tricks. It means you start expecting certain tasks to be handled by AI as a norm, and you plan your projects with that assumption. It means your mindset shifts from "Let's see if AI can do this" to "Here's how we do this now, with AI as a given." When your practice reaches that stage, you've gone from being merely curious to truly capable with AI, and that's a competitive edge. This transformation is ultimately about being more effective and innovative in serving your clients.

These principles encapsulate how to approach building an AI-enhanced practice thoughtfully. To ensure you can act on them, use the following questions as a starting point for crafting your own AI strategy blueprint.

Your AI Blueprint Starter Questions

Before you dive headlong into implementing AI, take some time to reflect and plan. Use these questions to guide your blueprint. They'll help to ensure you're covering the important bases and staying true to your goals:

- **Where do I spend the most time that clients don't see?** Identify those behind-the-scenes tasks (e.g., research, documentation, admin) that eat up hours. These are prime targets for AI-driven efficiency. If clients aren't directly valuing it, can you do it faster or automatically?

- **What elements of my deliverables could AI help clarify or extend?** Think about your client deliverables (reports, presentations, dashboards). Is there data that could be visualized better with AI assistance? Could an AI-generated brief or FAQ accompany them? Your answers reveal value-added uses of AI that enhance your deliverable process, not replace it.

- **Which client touchpoints could benefit from follow-up or reinforcement?** Map the client journey and look for places where clients would love more attention or information. Perhaps after a workshop, they'd enjoy a summary (AI can draft that). Or after project close, they might benefit from that custom chatbot or a periodic update. Find those opportunities where AI lets you be there for the client more consistently without a proportional increase in your effort.

- **Do I have systems in place to document and reuse my AI wins?** Plan for compounding benefits. If you discover a great way to use AI in proposals, how will you make sure you repeat it and that others (if any team members) do, too? Ensure you have a simple documentation habit. It can be as basic as a running Google Doc or a section in your notebook for AI learnings. This will help you build a knowledge hub over time and avoid forgetting the refinements that worked well.

As you answer these questions, you'll sketch the outline of your strategy. It might reveal, for example, that you spend 20 percent of your time on scheduling and follow-ups (ripe for a virtual assistant AI), or that clients often don't read your full report (so a summary bot could help), or that you haven't been saving any of the AI prompts you tried out (so you need a better system). That insight is a power that guides you to where AI can make a meaningful difference.

With answers in hand and all the preceding discussions, you're now equipped to draft a blueprint that is tailored to your practice. Finally, let's end the chapter with a reflection that zooms out a bit. So let's envision the change you want to see in your practice after embracing AI strategically.

From Tool to Transformation

Imagine looking back one year from now, after you've implemented your AI strategy blueprint. What might your practice look like? Envision this future.

You've built a custom GPT assistant that can answer frequently asked questions in your consulting niche, and it's become a trusted resource for both you and your clients. You've automated 80 percent of your proposal workflow, so what used to take days now takes hours, and your proposals are more personalized and insightful than ever. You've run multiple client workshops where AI helped you prepare materials and even debrief findings, leaving participants amazed at the depth of insight and responsiveness you delivered as a small firm.

But beyond these tangible outputs, something deeper has shifted. You've changed how you think and work. You no longer ask, "How can I fit this tool into my existing work?" Instead, you routinely ask, "What kind of new work or higher-level work does this tool enable me to do better?" In other words, you're rethinking old routines to actively invent new, better routines *around* AI. Perhaps you're taking on projects you used to shy away from, because now you have an AI-augmented method to tackle them. Or you're spending the extra time saved to connect more with clients, evolving from consultant to a true partner or coach for them.

This is the essence of transformation. That's strategy. That's reinvention. It's the difference between being one of the many curious onlookers of AI and being among the capable practitioners who harness it fully. And importantly, you've done it your way, aligning with your values, maintaining authenticity, and keeping the human touch at the center of your consulting practice.

In the end, that's what separates the curious from the capable. The curious try out new tools; the capable integrate them with purpose to elevate their whole practice. You have the courage to forge your own path with AI, the humility to keep learning, and the clarity to remain human-centric. With this blueprint, you're charting a new course for your consulting future, one where

technology and humanity work hand-in-hand to deliver extraordinary results. Welcome to the journey of becoming an AI-enhanced consultant in the
truest sense. Your next level awaits.

Part 2

Tools & Tactics

Once your mindset is aligned, it's time to get tactical. Part 2 brings AI down to earth—into the apps, workflows, and automations that can immediately elevate your consulting practice. These chapters are a field guide to what's out there and what's working, designed not to overwhelm but to orient. You don't need to master every tool; you just need to understand what's possible, where to start, and how to choose the right solutions for your clients, your services, and your style.

We'll explore everything from off-the-shelf copilots and meeting assistants to no-code automations, custom GPTs, and data-driven research tools. Real-world use cases and application strategies ground the discussion, always returning to the same question: "How does this improve the way I serve clients?" Whether you're automating proposals, enhancing discovery, or rethinking service delivery, this section is about building practical fluency so you can lead with confidence, experiment with intention, and put AI to work where it matters most.

CHAPTER FOUR

Out-of-the-Box AI Copilots for Consultants

Not every consultant needs to be a developer. But every consultant should know how to speak the language of AI copilots. In today's landscape, a new class of AI-powered "consultant's assistants" has emerged—tools that anyone can use with minimal technical setup. This chapter explores these accessible, out-of-the-box AI copilots and how they can augment your work. Used wisely, they offer immediate value by extending your capabilities, enhancing your outputs, and radically reducing time spent on routine tasks. In my own experience, I've witnessed such AI assistants in action, helping project managers save roughly 50 percent of their usual time creating and distributing weekly project status reports, completing what was a four-hour task in less than two. The key, as we'll see, is to adopt these tools in a human-first way, as creative amplifiers of your expertise, not replacements for it.

A New Class of Consultant Tools

Just as spreadsheets once transformed the business analyst, AI copilots are now reshaping the modern consultant's workflow. They're fast, intuitive, and critically, require no coding or data science background. Recent advances have made powerful AI as accessible as a web browser; anyone with an internet connection can now harness large language models to generate content and insights. This means that whether you're drafting a proposal, generating strategic options, or summarizing a seventy-page white paper, these tools compress time and expand your creative capacity. They allow you to offload mundane tasks and focus more on high-value human activities like building client relationships and solving ambiguous problems. Today's AI copilots come in a few core categories:

- **Conversational AI assistants.** These include general-purpose chat-based AI advisors that you can prompt in natural language, and are considered the "Swiss Army knives" of AI support—versatile for brainstorming, drafting, and Q&A. They typically run in a chat interface (often in your web browser or phone) and can handle a wide range of requests. Some advanced conversational assistants even support extremely large inputs (more than one million text tokens, or ~750 thousand words), letting you analyze multiple book-length documents or codebases in one go. The trade-off is that you must learn to craft clear prompts and instructions to get the best results.

- **Embedded productivity copilots.** These are AI assistants built directly into everyday tools like word processors, spreadsheets, email, and meeting software. They live where you already work—for example, suggesting formulas in a spreadsheet or drafting an email reply for you—so they require almost no new interface learning (low learning curve). They excel at contextual tasks, such as generating content that fits into an email thread, summarizing a video meeting recording, or creating a first-draft slide deck using your document outline. Because they're integrated, their power is in automating those tiny steps inside your workflow (e.g., *"summarize this thread"* or *"turn these bullets into a paragraph"*) without you needing to copy and paste into another app.

- **Research-enhanced assistants.** These are AI search companions that combine a chatbot with real-time web browsing or a knowledge base. Think of these as AI-powered research librarians; they can fetch up-to-date information from the internet and return answers with cited sources. For a consultant, this is incredibly useful when you need quick competitive intel, market statistics, or a thoroughly referenced explanation of an unfamiliar topic. Because they provide references, they help you double-check facts and integrate outside knowledge into your deliverables. Their learning curve is very low (if you know how to use a search engine, you can use these assistants), but the key skill is asking precise questions and vetting the sources returned.

- **Custom domain chatbots.** These are AI assistants tailored to a specific domain or even to your own organization's data. Without any coding, you can now create a custom chatbot that "knows" your proprietary content by ingesting a set of documents that you upload and having the AI answer questions based on those materials. These domain-specific copilots are like dedicated junior consultants; they won't perfectly understand nuance, but they can scale your expertise by handling FAQ-type queries and freeing you up for more complex work. Setting up a custom bot may require a bit of upfront effort (i.e., feeding it documents that are prepared for consumption by an LLM, integrating with your preferred AI service, and writing effective system instructions), but once built, it's as easy to chat with as any other assistant. The payoff is an AI that speaks your language, whether that's the jargon of your industry or the specifics of your firm's preferred methodologies.

Each of these copilot types offers unique strengths. A conversational assistant might generate a first draft of a client deliverable while a purpose- built AI tool could produce a finished PowerPoint deck or schedule your next meeting. A research assistant can gather third-party evidence to support your recommendations, and a custom bot can instantaneously answer a client's niche question by drawing on internal reports. The savvy consultant learns to mix and match these tools, selecting the right assistant for the job, much as a carpenter chooses between a hammer, screwdriver, or saw.

Given these and similar scenarios, it should be clear that there is no single "best" AI tool for all situations. The goal is not to use every new app out there, but to intentionally select the copilots that align with your needs and authentic consulting style. Some consultants thrive with a chat-based assistant always by their side; others lean on an integrated Excel plugin to crunch data. Be willing to experiment and find what works for you, rather than following the crowd. The tools will come and go, but the core use cases, such as speeding up analysis, sparking ideas, or automating grunt work, are timeless.

Extending Your Capabilities with AI Copilots

Many of these out-of-the-box copilots come with features that let you push their capabilities even further, still without writing a single line of code. Understanding these features can help you get the most out of whatever AI assistants you choose to use.

- **File uploads for context.** Many AI chatbots allow you to upload files (PDFs, Word docs, spreadsheets) or feed in large text to provide additional context to your prompt (like what you can do with a custom GPT, but for a single chat session rather than persisting for any number of sessions). This means you can ask questions about a lengthy document or have the AI summarize it for you. For example, you could drop in a fifty-page industry report and prompt the AI to extract the key trends and insights. If you provide relevant reference material, the AI's answers will be more specific and on target. This is a way of customizing the AI on the fly, giving it temporary expertise based on the files you provide.

- **Plugins and integrations.** Think of plugins as third-party apps for your AI copilot. With the right plugin, an AI assistant could, for instance, pull live data from a financial database, schedule a meeting on your calendar, run a calculation or even generate an image. For consultants, this enables powerful use cases. For example, you might use a data visualization plugin to create a quick chart for a presentation, or an Excel plugin to have the AI analyze a spreadsheet model. These integrations extend the AI's reach beyond text generation into taking actions or retrieving real-time information. As an example, one popular AI assistant's code analysis plugin can run Python code and analyze data you provide, essentially acting as a data analyst at your command. Another image-generation plugin can create custom graphics or concept art to spice up a proposal. While the range of available plugins will vary by platform (and some enterprise settings disable them for security), it's worth exploring what's available to supercharge your AI assistant when you need it.

- **Custom knowledge bases.** Perhaps the most game-changing feature for consultants is the ability to create a custom AI based on your own content. Like the custom domain chatbots described above, this typically involves uploading a collection of documents or providing a set of Q&A pairs to an existing service, like ChatGPT, that the AI can learn from. The result is a private chatbot that specializes in your chosen domain. Imagine having a personal AI trained on your firm's past proposals, research reports, and methodologies. It could instantly pull up relevant past insights or even draft new documents in your preferred style. This doesn't require sophisticated programming; many tools have user-friendly interfaces to add your data. The caution is that the AI will only be as good as the material you give it (garbage in, garbage out), and it won't truly *understand* the content beyond pattern-matching. But used appropriately, a custom knowledge AI can be like a junior analyst who has read everything you've ever written.

The Value of a Custom GPT

A quick story about how I've seen these things in action. A regulatory compliance consultant I worked with grew tired of answering the same client questions repeatedly. She compiled three years' worth of regulatory guidance PDFs and fed them into a custom AI assistant tool, effectively creating a "Compliance Q&A Bot." This assistant (named "STREG" to reflect state regulations) could instantly answer 90 percent of the routine questions clients or her staff analysts asked, citing the exact paragraph of the regulatory guideline. It became a billable differentiator for her practice; clients loved the rapid responses, and she could focus her own time on the trickier, nuanced questions.

In practice, treat these custom AI copilots like junior consultants on your team. They will be fast, tireless, and reasonably knowledgeable in their niche, but they won't get everything right. You still need to review their work and guide them, just as you would review an analyst's draft for accuracy. In other words, you can scale your thinking through such AI assistants, but you remain the quality control and strategic brain of the operation.

Real-World Use Cases Across Consulting Domains

How are consultants using out-of-the-box AI copilots today? Let's look at some consultant-friendly use cases, organized by specialty. These examples show practical ways to weave AI into your project work without any specialized IT setup:

- **Strategy consulting.** Use a generative assistant to brainstorm SWOT analysis factors or generate a list of possible strategic initiatives for a client's challenge. Employ a research assistant to quickly synthesize insights from several industry reports into a single summary. During early discovery, you might feed transcripts of stakeholder interviews into an AI to extract common themes or concerns, helping you map key drivers of the client's situation.

- **Operations consulting.** Leverage an AI copilot to draft process documentation and flowcharts after a process-mapping workshop. If you provide the basic steps, the AI can formalize them into a neat Standard Operating Procedure (SOP) document. For organizational design or project management, an AI can generate a RACI matrix (outlining who is Responsible, Accountable, Consulted, and Informed) from a simple list of team members and tasks. It's also handy for turning a rough procedure outline into a client-ready playbook with clear, concise instructions.

- **Marketing & brand consulting.** Use creative AI assistants to brainstorm value propositions and taglines, for example, by prompting for ten variations of a campaign slogan or brand statement. They're also useful for drafting social media copy or blog outlines in a given tone. Additionally, AI can simulate target customer personas by prompting with, *"How might a budget-conscious millennial respond to this product description?"* and get a plausible perspective to consider. This helps to test messaging and tone for different audiences.

- **Human capital consulting.** Deploy AI to help create role profiles and job descriptions by mixing and matching competencies. For instance, provide the AI with details of a role, and let it draft a first version of the job description or even suggest behavioral interview questions tailored to that role. If designing a training program, you can have the AI outline a curriculum or draft a scenario for roleplaying exercises. These uses save time on groundwork so you can focus on customization and nuance.

- **Technology consulting.** Have an AI assistant explain complex tech concepts in plain English for non-technical client stakeholders. For example, prompt with *"Explain the key differences between [software A] and [software B] in a one-page summary."* If you're reading through API documentation or technical specs, an AI with code or tech knowledge can summarize it for you or even generate integration diagrams. It's like having a translator between consultant-speak and developer-speak. Just remember to double check technical accuracy, as AIs can sometimes sound confident while being off base on details.

These are just a few illustrations, and the possibilities are endless. The rule of thumb is if a task takes more than thirty minutes and involves a lot of pattern recognition, summarization, or text generation, it's a candidate for an AI copilot to assist. For example, summarizing a long document, generating a first draft of anything, coming up with a list (risks, ideas, requirements), or analyzing free-form text data are all areas where AI helpers excel. By contrast, tasks that require delicate judgment, deep empathy, or heavy collaboration are usually better kept in human hands (though AI might still prepare background info for you). Successful consultants develop a sense for where AI is strong vs. where it struggles, and they delegate accordingly. One landmark study by a team at Harvard Business School called this finding the "jagged frontier" of AI capabilities. Staying just inside that frontier yields huge productivity boosts (nearly 40 percent improvement), while pushing AI to do things beyond its current abilities can actually hinder performance (a 19 percent drop when tasks fall outside AI's reliable zone). The key takeaway is to use AI boldly, but also wisely. Apply it where it can shine, and don't be afraid to roll up your sleeves when a task truly needs the human touch.

The Fifteen-Minute White Paper

I needed to draft a white paper on AI in one afternoon to meet a deadline. The first step I took was submitting a prompt using Gemini (Google's AI chatbot) to generate an outline of a white paper describing the benefits of AI in the insurance industry. I further prompted Gemini with the outline as a key input to create narrative drafts for each section. That was followed by an extensive rewrite which included a "humanizing" edit, deletion of superfluous and repetitive content, addition of "custom" content, and a verification of the data points cited by researching the source documents from which they were obtained. The initial drafting of the document—typically a two-or-three-day process—was completed in less than fifteen minutes. My rewrite, edits, and source verification consumed roughly ninety additional minutes. In total, Gemini helped me to reduce what was always a three-or-four-day process to less than two hours! (And, not to be overlooked, the white paper garnered more impressions and reactions than most of our firm's other posts on LinkedIn, and was, in fact, one of our most popular posts for the year.) This kind of rapid, AI-augmented workflow is becoming a competitive advantage. It doesn't replace the consultant's judgment on the final product, but it radically compresses the time needed to get to a solid draft.

Prompting Like a Top-Tier Consultant

When working with AI copilots, how you ask is as important as what you ask. In consulting, we're trained to frame problems and structure communication thoughtfully. Those same skills pay major dividends in getting quality output from AI. Crafting effective prompts is quickly becoming a core consulting skill, akin to asking a good interview question or writing a clear email to a client. In my experience, consultants who receive even elementary training on how to use AI (e.g., prompt engineering basics) perform significantly better with the tool than those who just wing it—so let's talk about how to prompt like a top-tier consultant.

- **Assign a role or perspective.** AI models are great at impersonating roles or adopting constraints. You'll get more relevant output if you set context up front. For example, *"Act as a risk management advisor for a healthcare client..."* or *"You are an executive coach* *speaking to a first-time manager... "* This helps the AI tailor its language and insights to the scenario, much as you would tailor your own approach for different audiences.

- **Frame the objective.** Clearly state what you're trying to achieve, for example by including *"I'm trying to help a client think through X"* or *"The goal is to generate options for Y"* in your prompt. This ensures the AI knows the purpose behind its task and can focus on information that serves that goal.

- **Define the desired format and scope.** If you need a table, an outline, or a bullet list, specify that. If you want a concise paragraph or a ten-slide outline, say so. For instance, *"Provide the output as a 2x2 matrix with brief explanations in each quadrant"* or *"Limit the response to one paragraph of no more than one hundred words."* The more you can shape the "container" of the answer, the less time you'll spend later reformatting it yourself.

- **Clarify the tone and style.** Just as we modulate our tone for different clients (formal vs. casual, academic vs. persuasive), you can prompt the AI with a tone, like *"Use a professional and direct tone, but keep it empathetic"* or *"Sound like a knowledgeable yet approachable mentor."* This can be especially helpful when drafting client-facing text that needs to align with a certain voice (even the client's own voice, if you have examples to expose to the AI).

- **Iterate and refine.** Think of the first AI output as a draft or an initial consultant's analysis. It's rarely perfect. Don't hesitate to ask follow-up questions or give feedback, such as, *"This portion is too generic, can you provide a more specific example?"* or *"Rewrite the recommendation in a more concise way."* You can layer on details; if the first result is on the right track but missing a factor, tell the AI to include it in a second attempt. This iterative back-and-forth is where a lot of the value emerges, much like refining an analysis through peer review. Some of the best outcomes come after several rounds of prompting, each time nudging the AI closer to the target.

To get you started, here's a sample prompt template that combines these best practices that you can adapt to many different situations:

> *Act as a [type of consultant]. Based on the following [context], produce a [output format] that [achieves goal]. Use a [desired tone] voice. Limit the response to [length or structure].*

For example, you might fill this in as: *"Act as a strategy consultant. Based on the attached notes from my client kickoff meeting, produce a bullet-point summary that highlights the client's primary concerns and goals. Use a confident and consultative voice. Limit the response to five bullet points."* After the AI responds, you would review those five bullet points, verify they align with your understanding, and perhaps ask it to elaborate on one or two that you find most critical.

Prompting *is* consulting. When you craft a great prompt, you're essentially doing what you do with a client: setting expectations, framing the problem, and guiding the conversation toward a valuable outcome. So approach it with the same thoughtfulness. Be specific. Be strategic. And remember that you are always in the driver's seat. The AI might be a savvy GPS, but you choose the destination and the route.

Finally, keep in mind the principle of *adaptability*. Templates like the example provided above are not perfect for every situation; they're more like training wheels. As you gain experience, you'll develop your own style of prompting that suits your brain and your workflows. Creating a personal checklist or set of go-to phrasing for prompts can also be used to maintain consistency. But don't become overly rigid; feel free to experiment with new phrasings and approaches. The field is evolving fast, and often you'll discover a new prompting trick by just playing around. Treat it as a creative dialogue between you and the AI.

Copilots for Every Engagement Phase

Another way to integrate AI assistance into your practice is to look at the consulting engagement lifecycle—before the project starts, the delivery phase, and after wrap-up—and insert copilots wherever they can add value.

AI tools shine in the moments around the core human interactions of consulting. Think of them as your support team for preparation and follow-through. Following are a few ideas to help you through each phase.

Pre-Engagement (Business Development & Preparation)

- **Prospect research.** Before a big pitch or initial client meeting, use an AI research assistant to gather industry trends, recent news about the client's market, or even summaries of the client's own press releases and financial reports. In minutes, you can arm yourself with insight that would take hours to manually compile. This background boosts your credibility and allows you to tailor your approach to the prospect's context.

- **Personalized outreach.** When reaching out to a potential client or stakeholder, an AI writing assistant can help draft emails that match the recipient's tone or address their specific pain points. For instance, you can provide notes about the prospect and let the AI draft a friendly yet professional introduction email. Just be sure to edit and add a personal touch; authenticity matters in relationship building, and people can sense a canned message. Use the AI's draft as a starting point to save time, then infuse it with your genuine voice.

- **Proposal scaffolding.** Kick-start proposal writing by asking an AI to generate a skeleton structure. Provide the context (client industry, project type, known challenges) and have it suggest sections or even sample text for each section. You might get a rough executive summary, a draft project timeline, and some risk considerations. Rather than starting from a blank page, you're editing and refining an AI-provided outline. This can ensure you don't overlook standard elements and can prompt you to consider aspects you might not have thought of (for example, the AI might include a "Change Management Implications" section, reminding you to address that angle). Again, you will inject your uniquely human nuance and specifics, but the AI can handle the boilerplate and structure.

During Engagement (Delivery Execution)

While you're actively working on a project, AI copilots can be like an ever-present analyst and admin support that tackles critical but often routine activities.

- **Real-time meeting support.** If you have an important meeting (especially virtual), consider using tools that generate live transcripts and summaries. Some video conference platforms now offer real-time AI note-taking that will produce a meeting summary and action items as soon as you conclude. This means you can focus on the conversation without worrying about scribbling notes. After the meeting, you instantly have a written recap to share with participants or to reference for yourself. For example, if a stakeholder couldn't attend, you can send them the AI-generated summary of key discussion points and decisions, ensuring everyone stays aligned.

- **Data analysis and modeling.** Mid-project, you often have data to crunch, such as survey results, financials, process metrics, etc. An AI with a code or spreadsheet plugin can help you analyze this data quickly. You could upload a CSV of survey responses and ask for common themes and statistics, or use an AI in your spreadsheet to suggest charts and run regressions. This augments your own analysis and can catch patterns you might miss. It's like having a tireless data analyst who works at the speed of software.

- **Content creation and editing.** As you develop deliverables (reports, slide decks, memos), AI is extremely handy for drafting sections or polishing text. If you've written a technical section that feels too dense, ask the AI to rewrite it in simpler terms. If you're stuck coming up with a catchy title for a "findings" slide, ask the AI for five options. It can also generate illustrative examples or analogies if you need to explain a concept (e.g., *"Provide an analogy to explain Agile project management to a non-technical client"*). During crunch time, having an AI to handle these smaller tasks can free you to focus on the storyline and recommendations.

Post-Engagement (Wrap-up & Value Sustainment)

When the formal project is done, consultants often have additional work ensuring the client gets lasting value. AI can help here too.

- **Project summary and handoff.** Generate executive summaries of your work that different audiences can digest. For instance, after a long project, you might need a technical summary for the IT team, a high-level summary for the C-suite, and a set of FAQs for end-users or employees affected by the project. Rather than writing each from scratch, use AI to repurpose your final report into these different formats. Feed in your detailed report and prompt it to output a one-page CEO brief, then a more detailed Q&A document. This can ensure consistency while meeting various communication needs.

- **Client learning resources.** To reinforce the value delivered, you might leave behind some educational materials, such as a guidebook or a set of tips for new processes implemented. AI can help draft these quickly. For example, after a change management project, ask the AI to produce a write-up that explains the change and the rationale, which the client can use going forward. Or have it create a glossary of key terms and concepts that were part of your project. Essentially, AI can package your consulting knowledge into reusable assets for the client, increasing the stickiness of your recommendations.

- **Relationship nurturing.** Even after an engagement, you want to keep the relationship warm. AI can assist in generating personalized follow-up emails at various intervals (e.g., one month and six months after the project) checking in on the client's progress and perhaps sharing a relevant article or insight. You supply a genuine sentiment (e.g., "I was thinking about your team as I saw this news..."), and let the AI flesh it out. This helps you stay in touch with many past clients without each message taking a lot of your time. Just double-check each one for authenticity; it should sound like *you*, not a robot. The personal trust you've built is paramount, so use AI to scale your outreach only if you maintain your genuine voice.

Here's what this might look like in action: A leadership development consultant integrated an AI copilot into his post-engagement routine. After each leadership workshop he conducted, he would feed the session transcripts into the AI and ask it to generate a concise recap for participants, along with three personalized action items for each attendee (based on what they had shared during sessions). He would then review and refine those action items. The result was a custom one-sheet for each participant that summarized what they learned and provided specific suggestions for improvement—a genuine value-add for the client that ensured the workshop learnings and follow-up activities would persist. This level of personalization would have been far too labor-intensive to do manually for every attendee, but with AI, it's feasible. A simple use case like this alone can justify your investment in learning new AI tools, as it can quickly differentiate your service and delight your clients.

Ethical and Professional Considerations

As suggested earlier, with great power comes great responsibility. As we integrate AI into our consulting toolkit, we must do so with judgment and discretion. Following are some key considerations to ensure AI enhances (and never undermines) your professional integrity:

- **Confidentiality.** Client data is sacrosanct. Many AI tools are cloud-based and feed data into large models that you don't control. Never upload sensitive client information into a public or consumer AI service unless you have explicit permission *and* a clear understanding of the tool's privacy policy. Remember that anything you input might be stored on servers outside your control. For instance, OpenAI's terms (for tools like ChatGPT) historically allowed the company to retain and review user inputs. There have been cautionary tales of employees pasting confidential data into AI tools and inadvertently leaking it. If you must analyze sensitive text, look for solutions that allow local processing or use an enterprise-grade AI with privacy guarantees. When in doubt, sanitize or anonymize data before using it with an AI; replace real names or figures with placeholders. It's better to be overly cautious than to have to explain a breach of confidentiality to a client.

- **Transparency and disclosure.** Maintain honesty about your use of AI, especially in deliverables. This doesn't mean you need to label every slide "co-written by AI," but consider the expectations set in your client contract or communications. If a substantial portion of a report was generated by an AI assistant, you might note in the methodology or even verbally that you used an AI tool for initial drafting or research. And just like the law firm client who expects the second-year associate to tackle the "grunt work" to avoid the partner's $750 hourly rate, most clients will appreciate that you're being efficient. Transparency builds trust. It shows you are not outsourcing judgment, only augmenting your productivity. On the flip side, if you try to hide AI usage and the client discovers an error or stylistic quirk that reveals it, that could erode trust. So, be upfront *always*. Explicitly state that "We used an AI-based analysis to support our findings here, and then we validated the results" when you've done so. This also reinforces that you're in control and using AI as a tool, not letting it run unchecked.

- **Verification of facts (not blind trust).** AI models do not guarantee truth. They can produce statements that sound authoritative but are completely false—the infamous "AI hallucinations," where an output has no basis in reality. As a consultant, your reputation rests on accuracy and sound advice. So always verify important facts or analytics that an AI provides. If an AI research assistant gives you a statistic or quote, check the cited source (if provided) or find a corroborating reference. If an AI drafts a legal or financial recommendation, *never* pass it on without your own expert review. These models are *not* subject matter experts and can make incorrect assumptions that draw embarrassingly bad conclusions. Think of AI outputs as a draft from a junior analyst who is extremely fast but somewhat unreliable. You would review a junior analyst's work carefully; do the same for AI. This is especially critical for numerical analyses, legal language, or any content that, if wrong, could have serious negative implications. Double-check calculations, and use AI's strengths (speed, language) to complement your strengths (expertise, critical thinking). In short, trust, but verify. Or better yet,

71

distrust *until* you verify when it comes to the need for factual accuracy.

- **Quality and professionalism.** Maintain your standards. AI can crank out ten pages of text in seconds, but quantity is not quality. Ensure that any AI-produced content you use meets the clarity, coherence, and relevance standards you would hold yourself to. Edit ruthlessly. Remove any hint of the generic fluff that AI often produces. Your clients hire *you* for insight and tailored advice, not boilerplate. Also be vigilant for biases or inappropriate content. AI systems learn from vast amounts of data that are riddled with biases. They might occasionally produce outputs that are culturally insensitive or just off-key for a professional setting. If you spot something like that, it's on you to fix it (and possibly to give feedback to the AI or avoid using that output). By filtering AI's contributions through your human judgment, you ensure that AI enhances rather than detracts from your professionalism.

- **Ethical use and client perception.** Consider how your use of AI will be perceived and whether it aligns with your client's expectations and values. For instance, some clients might be excited that you're leveraging cutting-edge tools, while others might be wary or old-school, expecting that all work is "hand-crafted." Gauge this and set expectations. Also, avoid using AI in ways that could be ethically gray, such as generating deliverables outside your scope of expertise and passing them off as your own expert opinion (a *major* no-no). AI can produce content on topics you're not trained in; that doesn't mean you should suddenly offer services in that area. Stay in your lane or collaborate with true experts. Use AI to support, not to masquerade. As a professional, it's better to say "That's beyond my current expertise" than to give a polished answer that you don't fully understand (even if AI helped create it). Humility and honesty go a long way.

Remember, your reputation depends on more than just efficiency. Using AI should never come at the cost of client trust. Speed and volume mean nothing if the work is wrong, misleading, or breaches confidentiality. Always pause to ask, "Is this enhancing my integrity and the client's trust, or just making me faster?" If it's the latter only, rethink the approach. The best consultants use AI to up their game, delivering analyses that are not only faster but also more insightful and well-rounded. The worst case would be to use AI as a crutch and become sloppy; that road leads to erosion of credibility. Fortunately, by keeping these considerations in mind, you'll steer clear of that trap and use AI in a way that *strengthens* your client relationships.

Building AI Habits, Not Hype

It's easy to be wowed by AI tools in a one-off demo. The real value, however, comes from consistent and deliberate use that build habits around AI in your daily routine. Just like any professional skill, small daily integrations of AI can compound into significant boosts to your effectiveness over time. The key is to move beyond sporadic experimentation and make AI a natural part of how you work. In that vein, here are a few tips to consider adopting:

- **Start your day with AI.** Kick off your morning by summarizing your overnight notes or prepping for the day using an AI assistant. For example, if you jot down thoughts or have a morning planning ritual, feed those into an AI and ask for a brief plan or summary, with something like, *"Here are my top three priorities for today and some notes; please generate a quick action plan."* This not only helps you to get organized, but it also trains the AI on your working context gradually. It's like a quick stand-up meeting with your digital assistant each morning.

- **Default to AI for first drafts.** Make it a habit that whenever you have to create a written artifact, whether it's an email, a proposal section, a slide blurb, or even a meeting agenda, you get an AI draft first. You don't have to use it verbatim, but you'll at least have something to react to. This can significantly reduce the intimidation some of us feel when facing a blank page. For instance, before writing a proposal intro from scratch, have the AI draft one and improve it. You might be surprised at just how effective this is, giving you a solid base that just needs your consultant's touch to shine.

- **Use AI as your research default.** Train yourself to ask an AI assistant questions before you do a manual web search. If your default browser homepage is an AI search assistant, you'll be reminded to do this. Often, the AI will give you a quick answer or at least guide your search by suggesting the right keywords and providing a synthesized snippet. For example, instead of combing through search results for "market size of X industry 2024," ask the AI. It might respond with *"According to [Source], the market size is Y,"* saving you time. Even though you should still click through to verify, you've accelerated the process. Over time, this habit can turn you into a much faster researcher.

- **Automate routine tasks with AI.** Identify one or two recurring tasks in your week that eat up time and see if you can partially automate them with AI. For example, if you spend fifteen minutes every Friday compiling a status update email to a client, start keeping a running bullet list of updates during the week and on Friday ask the AI to turn it into a well-structured email. Or if you frequently analyze similar datasets, create a standard AI prompt or script to do the first pass analysis for you. By creating these little AI workflows, you create leverage for yourself. The first few times might require some adjustment, but once it's set, you've essentially taught the AI a micro-skill that continues paying off.

These micro-habits ensure that AI isn't just a cool demo you try once, but a dependable extension of your working brain. Consistency is key; the more you incorporate AI, the more you'll discover its capabilities and quirks, and the better you'll get at using it. It's akin to using a spreadsheet—you might start with basic sums and over time learn advanced formulas and pivot tables, but only because you kept using it regularly. To spark your habit formation, here's a prompt for *you* to answer:

> *What is one fifteen-minute task I do almost every day that*
> *I could teach an AI to help me with tomorrow?*

Spend a moment to identify that task. It might be formatting slides, proofreading emails, tracking your time, or scanning news headlines for relevant bits of information. Challenge yourself to delegate it to an AI assistant in some form. Try it for a week and see what happens. You might win back an hour or two, but more importantly, you're training yourself to delegate to your digital team.

A final thought on habits is to remember to occasionally step back and evaluate. Not every habit will stick, and that's okay. Some AI uses may not prove as useful as you hoped. Feel free to drop them and focus on the ones that consistently give you a good ROI in time or insight. The goal is not to use AI everywhere for its own sake, but to use it intelligently where it adds value. Build the habits that make your consulting life easier, more productive, and maybe even more enjoyable (less drudgery, more creativity!). Leave the hype to others; you're building a sustainable, personalized system of working that leverages AI as an assistant, not a gimmick.

Tools Don't Replace Talent—They Extend It

At this point, you might be thinking of AI copilots as almost superhuman in their abilities, or conversely, worrying about relying on them too much. It's crucial to keep the right perspective. These tools are like an exoskeleton for your consulting muscles. They don't do the walking for you, but they help you lift heavier loads and run faster than you could alone. In other words, AI can extend your natural talents and effort, but the talent and effort are still yours.

Imagine you have an incredibly powerful calculator. It can compute insanely complex equations in milliseconds. Does having it automatically make you a great mathematician? Of course not! But it does allow a great mathematician to solve bigger problems faster. Similarly, an average consultant with AI will become more capable at handling analysis and routine tasks, but a *great* consultant with AI can reach new heights of creativity, insight, and impact. The differentiator is still the human consultant; your intuition, experience, and emotional intelligence guide the problem-solving process. The AI just gives you an extra boost along the way.

Many consultants who embrace AI report that rather than feeling replaced, they feel liberated. By automating the grind, they can spend more time on the aspects of consulting that technology can't do, like building trust, thinking strategically, and exercising creativity. You'll likely think, "I'm not faster because of AI. But I *am* freer. I get to spend more time on what I love, ask better questions, and draw far more compelling conclusions." (Better questions lead to better answers, after all.) This sentiment captures the essence of human-first consulting in the AI era—allowing machines to handle the busy-work so that humans can do the deep work.

It's worth noting, too, that using AI can be a catalyst for your own learning and creativity. When you see an AI generate multiple ideas in just a few seconds, you're exposed to thought patterns you might not have had. Some may be useless, for sure, but some are likely to be novel and intriguing. Those are the ones you build on. When an AI summarizes a complex document, you're essentially speed-reading with help, allowing you to consume more knowledge and make new connections faster. In a way, your AI copilot can be a brainstorming partner, a research assistant, and a tutor all in one. Use it to sharpen your thinking, not to blunt it. Don't just take the first answer it gives. Interrogate it, improve it, mix it with your perspective. The fusion of AI's breadth and speed with your depth and judgment is a potent combination.

Before we conclude, let's address a subtle fear many have—the fear that relying on AI might weaken your own skills, like a muscle that atrophies if not used. It *is* a valid concern; if you over-delegate your thinking, you might lose some agility. The solution is to use AI actively, not passively. Treat it as interactive; you're still solving the problem, just with a capable assistant. Stay intentional about occasionally doing things "the hard way" to keep your skills sharp. For example, if you regularly use AI to draft slides, draft them yourself occasionally to ensure you haven't lost your touch. Think of AI as a calculator again, where you still learn math fundamentals, but you use it to save time on the heavy lifting once you know what you're doing. If you maintain that mindset, AI will extend your capabilities without eroding them. In fact, it can push you to focus on higher-order skills involving interpretation, synthesis, and relationship-building that become even more potent when lower-order tasks are automated.

In summary, AI copilots are tools, and tools work at the direction of the craftsperson. You are the craftsperson. These particular tools happen to be astonishingly powerful, but it's your guiding hand and creative vision that give them purpose. Keep that orientation, and you'll never feel threatened by the tools; you'll feel empowered by them.

Small Tools, Big Leverage

You don't need to be a techno-wizard to start leveraging AI in your consulting practice. But you do need to be intentional. A little strategy in how you adopt these copilots will go a long way. To get started:

- **Choose one AI assistant to pilot.** Don't overwhelm yourself trying ten new tools at once. Pick one (perhaps a general conversational AI, or the built-in assistant in software you already use) and make it your focus to learn its ins and outs. As with any tool, mastery comes with use. Commit to using this one assistant consistently for a few weeks in your workflow.

- **Identify one use case per client engagement phase (before, during, after).** For example, *before*, use AI to research your client's industry; *during*, use AI to summarize weekly progress for the team; *after*, use AI to draft the project retrospective. By embedding the tool in each phase of your work, you'll see a variety of benefits and build confidence in its utility. It will also impress clients across the board, from sales conversations to delivery to wrap-up.

- **Develop one repeatable prompt or workflow that saves you at least thirty minutes a week.** Maybe it's a standard prompt for cleaning up meeting notes, or a template for generating risk registers from project plans. Refine it, test it, and then use it routinely. This becomes your personal "automation asset." Once you see that tangible time saving, you'll be motivated to find the next thing to delegate to AI.

After following these steps, step back and assess. You'll likely find that your productivity has bumped up, and perhaps more importantly, your mental energy is being spent more on thinking and less on slogging. That's the real win. From there, you can scale your use of AI systems by adding another tool to your toolkit, creating more custom prompts, integrating AI deeper into collaborative workflows. The idea isn't to work more (by all means, use the efficiency gains to work *less* or achieve a better work-life balance); the idea is to amplify your impact.

As you integrate AI copilots into your practice, keep the philosophy of this book in mind: human-centered, authentic, creative consulting. Let AI handle the grunt work, but put *people* first, including your clients and your own human touch. Use tools that resonate with your authentic style, so you remain genuine in how you deliver value. And perhaps most importantly, use AI as a catalyst for creativity. Try things that weren't possible before, venture into new analytical approaches, ask "What if?" more often because now you have a way to quickly test ideas. In doing so, you'll design a consulting practice as distinctive as your own values and vision.

In the end, the consultants who thrive with AI will be those who see it not as a threat or a shortcut, but as a partner in innovation. You're venturing into uncharted territory with a versatile copilot at your side and taking your clients with you. So start small, stay curious, and keep it human. Big leverage awaits.

◆

CHAPTER FIVE

Building Custom AI Assistants

Off-the-shelf AI tools are powerful, but custom AI assistants unlock a new level of relevance, efficiency, and differentiation for consultants. In this chapter, we explore how to build domain-specific AI assistants using no-code and low-code platforms. We'll demystify what it takes to create your own custom GPTs, chatbots, or automation workflows without writing a line of code, and show how even modest investments can yield major consulting leverage. Throughout, we'll keep a human-first perspective by demonstrating how to use AI to amplify your expertise, not replace it, and to free you to do more of what only humans can do.

Why Build a Custom AI Assistant?

When you create a custom assistant, you're encoding your expertise, preferences, and workflows into a scalable asset. It's like cloning a part of your brain, making it always available, always consistent, and immune to context-switching fatigue. Instead of a generic tool, you get a digital teammate trained in your way of working. Key benefits include:

- **Automating repetitive tasks.** Free yourself from rote work (scheduling, formatting, researching, etc.). Consultants can save many hours by using automation tools; that's time you can reinvest in high-value activities.

- **Scaling client education and support.** Your assistant can answer FAQs or guide clients through common processes on demand. This ensures no client question goes unanswered just because you're busy or asleep.

- **Delivering more consistent outputs.** Every deliverable or answer follows *your* established practices. The AI won't forget steps or skip details, ensuring quality control.

- **Differentiating your brand.** A custom AI can embody your unique frameworks and tone. It's a signature service that sets you apart. Rather than using the same off-the-shelf tools as everyone else, you're offering something authentically aligned with your consulting identity.

Perhaps most importantly, a well-designed assistant supercharges your human touch. By automating grunt work, you gain more time and headspace for relationship-building, creative problem-solving, and strategic thinking—the human elements that clients ultimately value most. In other words, the assistant handles the travel itinerary, but you still do the driving.

Consultants are increasingly capitalizing on this opportunity, with one of the highest AI adoption rates of any industry. And generative AI usage in professional services (which includes consulting) has doubled in the last year. The reason is clear; when done right, these assistants pay for themselves in time and value. Even simple automations have yielded big returns. For instance, a development team I managed found "at least 50% efficiency gains" by using an AI coding assistant. It's really just a matter of time before we see full adoption, where AI is simply another tool used to support improvements in the way we all work. The good news? Building a custom assistant has never been easier or more worthwhile, and embracing this trend can keep you competitive in a fast-evolving landscape.

A Custom GPT in Action

To provide one simple application, a human resources consultant built a custom GPT that guides HR managers through hiring reviews using her proprietary framework. The assistant walks managers step-by-step through evaluating job descriptions and interview processes for bias, essentially acting like an on-demand coach. She initially shared it with clients after engagements as a value-add resource. It was so effective that one client later licensed the assistant for their entire HR team, giving the consultant a new passive revenue stream and extending her impact firmwide. With that example in mind, here's how you can go about developing your own custom GPT:

Getting Started

Identify a valuable use case. Not every process needs an AI assistant. Success starts with picking the *right* use case. Begin by looking for patterns in your work by considering these questions:

- What questions do clients ask repeatedly? (e.g., "Can you explain how this framework works?")

- Which parts of your process are highly repeatable? (e.g., onboarding questionnaires, drafting similar reports.)

- Where do you spend time customizing (but not truly personalizing)? (e.g., tweaking standard templates or emails for each client.)

In other words, hunt for tasks that are common and time-consuming, but where a lot of the effort is doing similar things over and over. Those are prime candidates for an AI assistant.

Common use case categories. With respect to specific areas for which an AI assistant might prove most valuable to support consulting work, consider these:

- **Education.** Explaining frameworks, methodologies or jargon to clients in simple terms.

- **Diagnostics.** Intake forms, readiness assessments, or discovery surveys to gather info.

- **Process support.** Drafting standard operating procedures (SOPs), interview guides, or prep materials.

- **Follow-up tools.** Post-engagement check-ins, reminder bots, or clarification Q&A after a training.

- **Internal automation.** Note cleanup, report formatting, CRM data entry, or other internal busywork.

Evaluation framework. To evaluate your ideas, it helps to use a quick framework. For example, you might score a potential assistant idea on four dimensions:

- **Time saved.** How much time could this realistically save you per week?

- **Client impact.** Would this noticeably improve the client experience or results?

- **Repeatability.** Is this a process you can reuse in many different situations with minimal modification?

- **Ease of build.** Could you prototype a solution in under a day with available tools?

Focus on ideas that rank high on time saved and repeatability, have at least moderate client impact, and are feasible to build with minimal effort. In practice, that often means targeting the low-hanging fruit, like tasks you do frequently and similarly across projects. A Pareto analysis is useful here, where you identify the 20 percent of tasks that consume 80 percent of your time, and consider automating those first. Start with high-time, high-repeatability, moderate-ease wins. You can always tackle more complex use cases later, once you've got a quick win under your belt.

Assistant Types and How to Build Them

Now that you have a use case in mind, what kind of "assistant" are you actually building? Broadly, custom AI assistants for consultants come in various flavors, described in detail below.

Custom GPTs. These are like your own personal ChatGPT, tailored to your domain. OpenAI's ChatGPT Custom GPT builder (available with ChatGPT Plus, Team, Enterprise, and Pro accounts) allows you to create chat-based assistants that follow detailed instructions you provide and even leverage your own content. You can upload files (such as your slide decks, frameworks, or work samples) to provide additional context for the GPT. The result is an AI chatbot that speaks in *your* voice. You can interact with it yourself or share it via a public link with clients.

Building a custom GPT is surprisingly straightforward. For example, with OpenAI's interface you can simply log in to ChatGPT, click "Explore GPTs" in the sidebar, then hit "Create" to start a new custom bot. You'll be prompted to enter instructions describing the assistant's role and behavior, and you can upload any reference documents it should use. Essentially, you're giving it a brain dump of your expertise and guidelines. Then you chat with the GPT in a testing panel to refine its responses until you're satisfied. Finally, give it a name and save it. No coding required. In a matter of minutes, you have a shareable chatbot specialized in your domain.

To illustrate, a project management consultant might create a *"PMO Coach GPT"* by uploading his playbook for running projects and instructing the AI to answer questions as a seasoned project manager. He can then use this GPT to generate draft project plans or to have interactive Q&A sessions. If a client struggles with, say, risk management, he can share the GPT link so the client can ask it questions anytime. Given that ChatGPT is already the most widely used AI tool in many workplaces, leveraging it in a customized way is a natural next step.

Use custom GPTs when you need an interactive advisor that can converse, explain, or generate text in a specialized way. They're great for things like an "explain my methodology" bot, a role-play simulator (e.g., simulating a stakeholder for interview practice), or a client self-service Q&A tool. The strength of custom GPTs is their flexibility in understanding and generating language, powered by the latest large language models, with your custom spin on it.

No-code chatbots. No-code chatbot builders provide a visual way to design conversational agents. These can range from simple rule-based Q&A bots to AI-infused assistants that understand natural language. The idea is you drag and drop blocks representing messages, questions, or actions to create a dialogue flow. Platforms like Voiceflow, Tiledesk, and others have user-friendly interfaces that require no programming experience. You define how the chatbot should greet a user, what options or questions to present, how to respond to certain inputs, and so on—all through an easy-to-use editor. Many of these tools can integrate LLMs in the backend, so you can combine scripted responses with AI-generated ones for flexibility.

Using Voiceflow as an example, you can start from a template (say, a FAQ bot) or from scratch and visually map out the conversation steps. Instructional logic (e.g., "If user says X, do Y") is handled through the interface. You might create intents or keywords that the AI will recognize (for instance, if a user's message contains "pricing," the bot will trigger a response with pricing info). It's very much a design exercise, not a coding one. Voiceflow's drag-and-drop editor makes it easy to prototype and test the conversation as you go. When you're happy, you can deploy the chatbot to your website, Slack, WhatsApp, or other channels with one click. Many platforms allow you to try them out at low or no cost.

These shine for structured interactions like lead qualification, onboarding sequences, or FAQs. For example, a marketing consultant could build a lead-gen chatbot on her website that greets visitors, asks a few questions about their needs and budget, then either answers basic queries or directs hot leads to schedule a call. Or a human resources consultant might deploy a chatbot that helps employees navigate common HR policies (*"How do I request PTO?"*). The bot handles the repetitive Q&A, so the consultant only intervenes in complex cases. No-code bots are ideal if you want something user-facing on your site or social media that provides interactive support or data collection without requiring your constant attention. They deliver a consistent experience to every user. Plus, having a slick chatbot can differentiate your digital presence by signaling that you're tech-savvy and responsive.

To provide a relatable example, suppose you'd like to reinforce training for new managers between workshop sessions you facilitate. Using Voiceflow, you could build an onboarding chatbot that lives on your client's intranet. New managers can chat with the bot to practice difficult conversations, get reminders of key points you covered, and even receive tips on handling common management scenarios (e.g., conflict resolution, managing up, mobilizing team members, etc.). You might instruct the bot to use a friendly tone and integrate OpenAI to allow some free-form Q&A. Your clients effectively gain an on-demand coach, not one that's available just during scheduled workshops, adding value to your work while freeing you from repeatedly answering the same basic questions.

Low-code workflow automations. Low-code automation platforms let you create multi-step workflows that connect various apps and apply AI at certain steps. Think of it as building a custom pipeline where you define a trigger (e.g., "a new client email arrives" or "form submitted" or "meeting ended") and then a series of actions that should happen (e.g., "send the email text to GPT-5 for summary, then save that summary to Notion, then alert me on Slack"). These platforms provide a visual editor where you can chain steps together and integrate with thousands of apps. You don't write the code; you configure the steps using pre-built modules and simple logic.

Popular tools in this category include Zapier, Make, and n8n. Zapier is known for its vast library of app integrations and very easy interface—often a good choice for straightforward "if X, then Y" processes. Make offers a bit more flexibility and a visual canvas to map more complex flows (and it's friendlier to custom API calls), while n8n is an open-source alternative that you can self-host and extend with code if needed. All three can work with AI; they have modules or integrations for services like OpenAI, allowing you to plug in an AI action wherever you need one. For example, Zapier has a built-in OpenAI step where you just paste your prompt and it will return the AI's result as the next step's input—no need to call the API yourself.

Use these for behind-the-scenes automations that glue together different tools in your stack, especially when there's a clear sequence of tasks. They excel at things like automatically transcribing voice and generating a summary that is then emailed after a client call; or taking form inputs and producing a customized document about which you're instantly alerted once completed. Essentially, any time you find yourself doing a routine process across multiple apps, consider automating it. For instance, an operations consultant might set up a scenario using n8n such that when a client fills out a Typeform questionnaire, the answers are sent to OpenAI to draft a first version of a procedures document, then that draft is saved into a Google Drive or Notion folder and an email notification is sent to both the consultant and client. All this happens in minutes without human intervention.

Low-code tools are about extending your reach and ensuring nothing falls through the cracks. They work in the background, handling the busywork while you focus on the higher-level consulting. The great part is you can start small (e.g., automate just one step of your process) and gradually build from there. Many consultants using Zapier start with a simple "Zap"—say, when they add a tag to a contact in a CRM, trigger an AI to draft a follow-up email—and then iterate from there. Over time, you might assemble a whole sequence that runs by itself. It's like having an invisible assistant who never forgets a step.

In practice, let's say you spend hours producing standard operating procedures (SOPs) for your clients as part of your service offering. You could implement an automation using n8n that enables clients to fill out a Typeform with their process details. The n8n workflow can then parse the responses and feed them into a ChatGPT prompt (via API) to generate a draft SOP for you to review (*always* keep a human in the loop for client deliverables!). You then trigger the workflow to send the draft downstream for formatting and saving in a client's Notion or other workspace for review. What used to take hours of tedious writing might be reduced to just fifteen or twenty minutes! You'll get a *dramatic* efficiency boost, and clients will get their documentation faster.

In short, custom GPTs, chatbots, and automated workflows are complementary tools. You might use one or all three depending on the need. A custom GPT could be your content generator and expert Q&A, a chatbot could be your friendly front-end for client interaction, and a Zapier or n8n workflow could tie it all together with your other systems. The key is to choose the right approach for the job.

Matching Tools to Use Cases

To summarize some of the options we've mentioned, here's a quick comparison of popular platforms for building AI assistants, and what each is best suited for:

Platform	Best For	Tech Skill	Cost (2025)
ChatGPT Custom (OpenAI)	Advisory bots, FAQ assistants (chat-based)	Low: No code, just writing instructions	~$20/mo (ChatGPT Plus)
Voiceflow	Dialog-based onboarding or website chatbots	Low: No code, visual flow design	Freemium (paid tiers for high usage)
n8n	Workflow automation with flexibility	Medium: Low-code, logic configuration	Free (self-host) + paid cloud options starting at $20/mo
Zapier	Business task automation with many integrations	Low/Med: No code, very user-friendly	~$20+/mo (depends on usage)
Make	Visual API workflows and multi-step processes	Medium: Low-code, more advanced scenarios	Freemium (paid for higher operations)

Tool selection guidance. As you can see, you don't need to be a programmer to get started with any of these. The best tool is the one that integrates with your current stack and fits your comfort level. Don't let shiny features lure you into a complex platform if a simpler one will do the job. It's usually wise to start with the tool that feels most accessible to you and that plays nicely with the software you already use (Google Docs, Slack, CRM, etc.).

One of the secrets to success in building custom assistants is matching the right tool to the right use case. Here's a quick pairing guide for common consulting scenarios and the platform that might fit best (not a recommendation; just for illustration):

Use Case	Best Tool	Why It Fits
Explain your frameworks or methodology post-engagement (interactive Q&A)	**ChatGPT Custom**	Can be pre-loaded with your documents and handle open-ended questions.
Website intake or FAQ bot for prospects	**Voiceflow (or similar)**	Easy to embed on a site, supports rich dialogues with a polished UX.
Generate documents like SOPs or reports from client input	**Make + OpenAI**	Handles multi-step workflows from form input to AI generation to output formatting.
Generate and deliver a daily email digest	**n8n + OpenAI**	Eliminates need to sift through dozens of daily emails to provide only pertinent or actionable content in a single email at the end of each day.
Update CRM and send Slack/email after a client call (triggered tasks)	**Zapier**	Event-driven, great for connecting business apps and sending updates without manual effort.

While these are reasonable pairings, keep in mind that the best solution typically involves integrating with your existing workflow, and some tools work better with others, and decisions around tool selection should always consider your existing tool stack. For example, if you already use Slack with clients, select a platform with pre-built Slack integration. If your deliverables live in Google Docs, make sure you can easily access your Google account with the tool you choose. Always consider practicality over novelty.

Consulting-Specific Use Cases

To spark some inspiration, below are common ways consultants are deploying custom AI assistants in their practices:

- **Custom GPTs.** Easy to build, custom GPTs can add significant value to your practice by making your expertise available 24/7 with very little effort.

- *Interactive methodology coach.* Clients or juniors can ask the GPT to explain your models or get guidance, just as they would ask you, the expert.

- *Stakeholder interview simulator.* Practice Q&A with an AI that plays the role of a tough client or executive, based on profiles you've provided.

- *Post-engagement coach.* After you finish a project, give clients a GPT trained on the project insights and next-step recommendations. It can answer their questions as they implement your advice, extending support without you being on call.

- **No-code chatbots.** The next step in your AI evolution might involve a no-code chatbot to handle routine activities that previously required your direct intervention.

  - *Website lead qualifier.* An embedded chat on your site that greets visitors with a brief query, such as *"Hi, what are you looking to improve?"* and funnels qualified leads to book a meeting, while politely answering others' basic questions.

  - *ROI calculator bot.* An interactive tool that asks a series of questions and then calculates potential ROI or benefits (with a bit of math or AI reasoning under the hood) for your service, giving prospects instant value.

  - *FAQ & resources bot.* A chatbot that lives on your client portal or Slack, which team members can ask *"How do I do X?"* and it responds with answers drawn from documentation or your knowledge base.

- **Low-code automation.** Taking things a step further, you might dabble with a low-code process orchestration platform to automate more complex activities and save yourself from the drudgery of important but often tedious administrative and related work.

  - *Project summary.* At the end of each work week, automatically aggregate outputs from multiple project management tools (e.g., Jira, Smartsheet, Confluence) and generate summaries and key action items for each project. The assistant parses by project, emails the summary to relevant stakeholders, and updates your project tracker.

o *Proposal generator.* You fill out a short form (client name, their industry, project type, etc.) and an automation workflow produces a first draft proposal or engagement letter using your templates, ready for you to polish.

o *Continuous insight monitor.* Set up a process that periodically pulls data (market stats, social media mentions, whatever's relevant), uses AI to analyze trends or sentiment, and delivers a monthly insights report to you or your client. This turns labor-intensive research into an automated service.

These examples barely scratch the surface. The beauty of no-code and low-code tools is how much you can customize a solution to fit *your* niche needs. As you experiment, you'll likely uncover many more opportunities where an assistant could add value.

Building Your First Assistant

By now, you might be thinking, "Alright, I have an idea. What do I actually need to build this?" Fortunately, the requirements are more about time and clarity than resources. Here's a basic checklist:

- **A clear use case.** One that likely saves you (or your client) at least thirty minutes a week once automated. Start with a small but meaningful problem to solve.

- **A preferred tool in mind.** Based on the sections above, choose a platform that you feel comfortable with. (If in doubt, start with the simplest option, like ChatGPT Custom or a pre-built n8n template.)

- **Sample content or prompts.** Gather any existing material you can use to train or instruct the assistant. This could be sample prompts you've written, documents to upload to a GPT, or a list of FAQs for a chatbot.

- **An hour or two to experiment.** Block out an afternoon or weekend slot to just play around. Treat it as a sandbox session, not a critical project.

That's really it. You don't need a development team or a big budget. You can easily build the first prototype in a couple of hours. The key ingredient is curiosity and a willingness to experiment and learn. Remember to approach this with a bit of humility; your first version won't be perfect, and that's okay. You're exploring new territory (which, incidentally, is what being a consultant often entails).

To give your custom GPT a solid foundation, you'll want to craft a clear system prompt or instructions. Here's a starter template you can adapt:

> *You are a consultant specializing in [your domain]. You help users with [types of tasks or questions]. Respond in a tone that is [friendly/professional or other desired tone] and prioritize [certain values, e.g., clarity or empathy]. Base your responses on the reference documents provided (do not make up answers not supported by the documents). Do not guess; if you are unsure, ask the user for clarification. Keep responses concise and actionable.*

This kind of prompt sets the stage for your GPT-powered assistant, telling it who it is, what it should do, and how to behave. You would, of course, fill in the specifics for your context and attach any relevant documents. Many custom GPT builders (like OpenAI's) allow you to input this as the system message or instructions that guide every response.

Your first assistant doesn't need to be perfect. But it does need to be used. Focus on making something useful for yourself, even if it's rough around the edges. You can always refine it. The worst outcome is spending days building an elaborate tool that you never actually integrate into your work. So build for your own needs first. If it reliably saves you time or effort, then you can consider sharing it with clients.

From Prototype to Practice

Once you've tested an assistant internally and it's working decently, you might wonder if it could be client-facing. Many great consulting offerings start as internal hacks or aids that proved their value. Turning your assistant outward can amplify its benefits (and even create new revenue streams or service offerings). Here are a few questions to evaluate that move:

- Would a client benefit from using this tool directly? If yes, in what way? (e.g., would it save them time or provide insight when you're not around?)

- Could this extend support after a project ends? For example, a chatbot that continues to answer questions for three months post-engagement can increase the longevity of your impact.

- Would clients pay for it (or pay *more* because of it)? Perhaps it's a premium add-on, or maybe just a differentiator that helps you win proposals. Even if not paid directly, a valuable assistant could tip the scales in a competitive bid.

If the answers indicate promise, you can start exploring how to package it. Sometimes the assistant is a free value-add to delight clients; other times it might be an upsell ("For an additional $2,500, I'll provide a custom AI assistant trained on our project that your team can consult for the next six months."). Be sure to consider maintenance and boundaries; you may need to update the assistant's knowledge or handle edge cases if clients rely on it.

For example, a leadership development firm might create a custom GPT trained on their coaching curriculum and case studies. During engagements, clients would be given access to it for between-session support. They might prompt the GPT with, *"How might I apply concept X in my situation with my team?"* and get guided reflections drawn from the firm's teachings. The clients would benefit greatly from having on-demand advice and are more likely to stay engaged. The overall quality of your engagements will improve as clients could be better prepared during regular interactions.

This is just one way an assistant can enhance your core service rather than replace it. We humans still do the high-touch work, but the AI assistant reinforces and scales our impact.

Build vs. Buy vs. License

As you consider implementing an AI assistant, there are a few paths to consider: build your own (what we've focused on here), buy or use an existing solution, or license a proven tool from elsewhere. Each has merits, which we'll explore below.

- **Build it.** If you want full control and reusability, this is what we've been discussing so far—using no/low-code tools to craft something tailor-made. It's great for differentiating yourself and you can refine it continuously. It requires some time investment, but you'll learn a lot along the way *and* own the result.

- **Buy a template or platform.** If you value simplicity and speed, consider buying something pre-built. Maybe someone has already created a chatbot template for consultants (many workflow platforms maintain large libraries of user-developed workflows, some of which are free, while some require license fees), or maybe there's an instantly available "AI proposal writer" that reflects nuances specific to your particular domain or discipline. Purchasing or subscribing to those can be quick to deploy, though you might sacrifice some customization, and of course someone else owns the intellectual property.

- **License a proven tool.** If you need scale and credibility fast, license a proven tool. For example, look for proven assessment chatbots for your industry that you could license and (potentially) brand as part of your offering. This can impress clients (since it's presumably a validated solution) and you don't have to reinvent the wheel. However, it might be costly and less unique to you and, again, you don't own it.

There's no one right choice, and you can mix approaches. You might build one assistant yourself and for another need, license a different one. The takeaway is that you don't need to productize your practice overnight. Start small. Try making just one part of your process reusable, like an automated diagnostic quiz that you previously did manually.

The key is not to get overwhelmed by possibilities. Pick one use case and one tool to start. It's better to have a simple assistant in action than a grand idea that never leaves the paper. Every automated workflow or chatbot you create is a learning experience. Over time, these can compound into a significant ecosystem of tools supporting your work, but it happens one step at a time.

Small Builds, Big Returns

You don't need a development team or a computer science degree to create a useful AI assistant. You need a clear problem, a good prompt, some validated content, and a few hours of focused play. The barrier to entry is low, but the potential upside is significant. Whether you're building internal tools to streamline your workflow or client-facing assets to enhance your service, custom AI assistants offer you a path to:

- **Consistency.** They perform tasks the same way every time according to your guidelines, which means less room for human error and more reliable quality in your outputs. For example, your report drafting assistant will always include all sections and follow your style guide with no missteps or forgotten sections.

- **Scale.** They allow you to serve more clients or deliver more value *without* a linear increase in effort. It's like multiplying yourself. If a client onboarding bot answers twenty common questions a month, that's twenty fewer emails you personally write, enabling you to take on additional work or simply reclaim that time.

- **Differentiation.** They embody your unique expertise and style, setting you apart in a crowded market. Using off-the-shelf tools might boost efficiency, but building your own assistant gives you ownership over a proprietary asset. Clients will remember that *you* were the one with that amazing bot or workflow, differentiating yourself from those who didn't. It demonstrates innovation and adds to your brand story as a forward-thinking advisor.

In the end, it's not about building tech—it's about building leverage. The technology is a means to amplify your impact and create more value for your clients and yourself. By delegating repetitive or structured tasks to an assistant, you free yourself to focus on higher-level consulting work—the creative, strategic, empathic work that AI *can't* do. That's where you truly shine, and now you'll have more room to do it.

Start Here: A Quick Action Plan

If you're ready to dip your toes into the world of custom AI assistants, here's a quick action plan to get you started:

- **Identify one high-repeat, moderate-effort task** in your workflow that frustrates you or eats up time.

- **Choose a platform** that fits that task (chatbot, GPT, or automation—use the tables above as a guide).

- **Draft a few core prompts or sample interactions** you'd want the assistant to handle. (This helps clarify how it should behave or what it should say.)

- **Build a small prototype.** Keep it super simple and test it on yourself. For instance, use your new GPT to draft a section of a report and see if it's helpful.

- **Iterate or toss.** If it shows promise, refine it; if not, no harm done—adjust the idea and try again.

- **Share it with a colleague or friendly client** for feedback once it's stable. Sometimes others will spot immediate uses or improvements.

By following these steps, you'll move from concept to something tangible. The first build is the hardest; it only gets easier from there as you learn.

Finally, keep in mind that this is a journey. You are essentially co-creating with AI to develop tools that support your unique consulting practice. Stay curious and brave. The fact that there is no one-size template means you have the freedom to invent what consulting augmentation looks like for you. As we emphasized early on, use AI as a creative, enabling force to design a practice as distinctive as your own values and vision. The consultants who get the best ROI from these technologies approach them as genuine collaborators—partners in innovation, and not simply time savers or efficiency boosters.

CHAPTER SIX

◆

Automation for Consultants

A utomation can not only save significant amounts of time, it can create *space*. In a consulting practice, "space" means more capacity for strategic thinking, deeper client relationships, and creative problem-solving. This chapter explores how consultants can use automation to streamline internal operations, enhance client service delivery, and free up energy for the human-centric work that matters most. From no-code tools like Zapier and Make, to integration platforms like n8n and even AI assistants—which we've previously introduced—we'll walk through real-world applications that generate tangible returns. The goal isn't to turn consulting into a robotic process, it's to remove friction and amplify your impact.

This isn't about replacing people; it's about removing friction. Leading firms emphasize that AI and automation are simply tools to enhance decision-making, not substitutes for human expertise. In fact, the consulting industry is increasingly turning to automation to reduce tedious tasks and speed up delivery to meet client demands. It's no stretch to say that consultants who want to remain competitive need to automate significant portions of their business *now*. Used thoughtfully, automation augments your consulting practice, handling the busywork in the background so you can focus on insight, trust-building, and innovation.

Why Automate?

In addition to dispensing advice, the best consultants design *experiences*. Automation helps make those client experiences more consistent, timely, and scalable. It turns manual tasks into background processes and transforms "I'll get to that later" into "already done." By significantly streamlining repetitive and time-consuming tasks, automation allows consultants to redirect focus toward high-value activities like strategic planning and deep client engagement. It can surface insights faster, ensure nothing falls through the cracks, and elevate professionalism at every touchpoint. And importantly, it does all of this without replacing the human judgment and creativity that consultants uniquely provide. Instead, automating routine processes supports you in delivering your expertise more effectively.

Benefits

The benefits of automation are many; following are just a few that can profoundly impact the productivity of even average consultants by, in a sense, "democratizing" efficiency.

- **Faster follow-up and task closure.** Automated workflows ensure that responses and routine tasks happen immediately, without waiting for your manual action. Prompt follow-up—for example, an instant thank-you email or scheduling link after a lead submits an inquiry—shows reliability and responsiveness. Clients today expect faster turnaround, and meeting those expectations helps maintain strong relationships. When mundane to-dos close out quickly, projects keep momentum and you free up mental space.

- **More professional client touchpoints.** Automation adds polish and consistency to how you interact with clients. Standardized email templates, reminders, and updates can be sent automatically at just the right moments, so every client gets the same high level of service. Ensuring tasks are performed consistently and on-time builds trust, since clients experience a dependable, well-organized process. These reliable, timely touchpoints make clients feel valued and show that you're on top of details.

- **Greater focus on strategic thinking.** Every hour not spent transcribing meeting notes or updating spreadsheets is an hour gained for analysis, problem-solving, and big-picture planning. By automating low-value tasks (data entry, scheduling, simple reports), consultants can devote more energy to complex work that requires human insight. In other words, automation shifts your time from *working in* the business to *working on* the business—honing strategy, coaching clients, and developing innovative solutions that truly move the needle.

- **Reduced error and rework.** Even experts make mistakes when copying and pasting text or performing even simple calculations. Automations execute processes the same way each time, minimizing the risk of human error. For instance, if your CRM automatically populates a report using a defined workflow, there's less chance of information being entered incorrectly or a detail being overlooked. This improved accuracy maintains quality and prevents the need to fix avoidable errors. Fewer mistakes mean higher credibility and less firefighting.

- **Consistency across engagements.** With well-designed automation, your practice runs like a well-oiled machine. Key steps happen reliably for every client; every proposal uses the latest template, every project onboarding follows the same checklist, every monthly report goes out on schedule. This consistency helps ensure no step is missed and every client receives the same high standard of service. It also reinforces your brand; clients come to know exactly what to expect when working with you, boosting credibility, confidence, and satisfaction.

- **Scaling your impact.** Ultimately, smart automation lets a solo consultant or small team achieve more with less. You can handle more clients or bigger projects without suffering missteps, because many tasks simply run at all hours without your involvement. Routine processes keep running even while you're sleeping or busy, dramatically increasing your productive capacity. This scalability is crucial if you're looking to grow your practice; it enables expansion without sacrificing quality of work. In short, automation helps your business scale up while *you* stay focused on the high-value work that truly requires your expertise.

To illustrate the power of automation, consider just how transformative these benefits could be. Let's say you build a simple Zapier workflow for new client inquiries that handles several steps automatically, including triggering a personalized welcome email sequence, updating your CRM with the lead's info, scheduling a discovery call via Calendly, and generating a draft proposal document—completing this otherwise routine (and often tedious) work within *minutes* of the inquiry. What used to take the better part of a day (and may slip through the cracks when you get busy) can happen in about ninety seconds while you go about your other work. The immediate, professional response will likely impress prospective clients and set a tone of efficiency before work even begins. More importantly, by offloading those onboarding logistics, you'll reclaim hours each week to spend on strategic research and one-on-one conversations. This scenario reveals that every task you automate predictably saves time, and also frees up energy to be reinvested into more valuable client-facing activities.

Where to Start

Faced with the possibilities of automation, a common question is "Where do I even begin?" The answer is to start with a simple audit of your current work. Take stock of your repetitive tasks and pain points before you dive into tools. Spend a week being consciously aware of what you're doing. Jot down tasks that you consider tedious or mundane. These will reveal prime candidates for automation. Ask yourself:

- **What tasks do I repeat every week?** Routine reports, weekly meeting decks, status updates, timesheets, etc. Recurring tasks, especially those that consume a lot of time in aggregate, are low-hanging fruit for automation.

- **Where am I juggling information between tools?** Do you routinely copy text from emails to put into Excel, or move Excel graphs into PowerPoint presentations? If you're regularly moving information between applications or entering data in multiple places, redesigning those activities through integration or automation can yield substantial time savings.

- **What do I postpone because it's annoying, not hard?** Perhaps updating your CRM, organizing notes, or sending follow-ups gets pushed to the end of the day (or week) because you dread it. These tasks aren't intellectually challenging, they're just tedious. They're perfect to automate because you won't miss doing them and your schedule will gain time to get into the flow of consulting rather than the tedium of administrivia.

From this audit, you'll likely notice patterns emerging regarding activities that are the biggest time-sucks. You're probably noting that many administrative tasks fall into just a few larger categories of automatable work, including:

- **Client onboarding.** This includes preparing proposals or contracts, gathering initial client information, sending welcome or orientation emails, and onboarding questionnaires. These can often be templatized and triggered to send with minimal manual input.

- **Document generation.** Presentation decks, proposals, or engagement reports where a lot of the content is standard or pulled from templates all fall into this category. For example, you might consider generating a first draft of a proposal or a meeting recap document by merging client-specific details into a pre-formatted template using AI tools.

- **Meeting management.** Scheduling meetings and sending reminders, setting agendas, recording or transcribing discussions, and following up with notes and action items often consume an inordinate amount of time. Automating calendar invites or having an AI meeting assistant prepare and distribute a summary after a call reduces the tedium associated with such activities.

- **CRM and administrative work.** The need to regularly update contact records, log meeting notes, track deal stages, distribute project status boards, issue invoices, and perform other administrative updates tends to infringe on large portions of the workday. These activities are critical for operations but ripe for automation, so they stay consistently updated without consuming too much time.

- **Internal ops.** Task assignments, project timeline updates, internal status reports, time tracking, expense logging, and related operational activities consume considerable time as well. Often, internal operations work can be streamlined with simple workflows (for instance, automatically creating a task in your project tracking software when a client emails a request).

Chances are your audit will identify at least a couple of tasks in each category. But how do you prioritize what to automate first? Not all tasks are equal in value. This is where a simple framework can help. Think of a two-by-two matrix plotting frequency (how often a task occurs) against friction (how painful or inefficient the task is) to establish automation priorities. High-frequency, high-friction tasks sit in the top-right of this matrix; they're the ideal starting points for automation, because they're regularly eating up time *and* they consistently cause frustration or delays. By tackling those first, you'll free up maximum time and alleviate the biggest headaches.

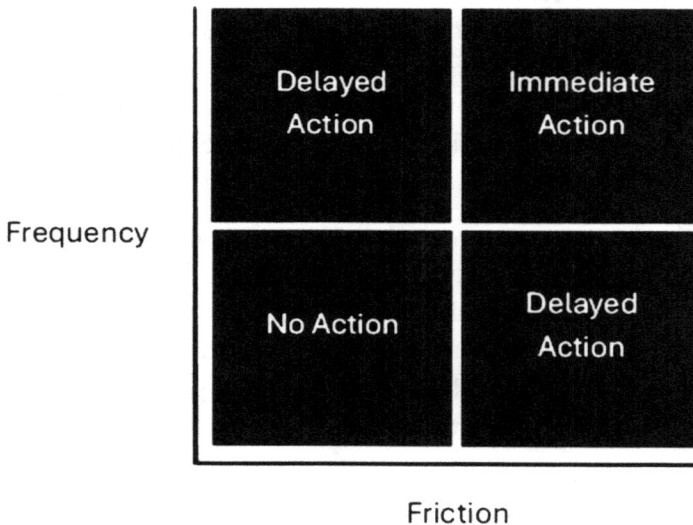

Delayed Action	Immediate Action
No Action	Delayed Action

Frequency

Friction

Automation Prioritization Matrix

Consider the dreaded weekly timesheet that many consultants are still (remarkably) required to complete. You could spend as much as thirty minutes filling out a timesheet and ensuring others in your charge do the same, requiring reminders to those who drop the ball and forget. This is a great example of a high-frequency (every week by every project participant) and high-friction (everyone finds it annoying, and it is often delayed)—in other words, a perfect automation candidate. A simple solution is an AI agent designed to pull calendar entries and match them with project codes and then generate each person's time sheet in a .csv file that could easily be uploaded into most time-tracking systems. The resulting low-touch timesheet process would yield more consistent and timely entries and far fewer weekly headaches.

Not every automation will yield such a clear win, but by focusing on tasks that are both common and cumbersome, you're more likely to see immediate benefits. On the other hand, if a task is low-frequency or only mildly annoying, it might not be worth automating right away. Save those for later. Start where automation will noticeably lighten your load. This way, you build confidence (and some quick time-savings) that can fuel more ambitious automation projects later.

Tools of the Trade

Fortunately, you don't need to be a programmer to automate many consulting tasks. A wealth of no-code and low-code tools make automation accessible to anyone willing to do a bit of drag-and-drop or template configuration. Each tool has its own strengths, so it's about finding the ones that fit your needs (and your comfort level). Here are a few popular platforms and what they're best used for:

- **Zapier.** A pioneer in no-code automation, supporting quick connections across over 7,000+ apps (everything from Gmail to Salesforce to Slack). It's best suited for simple, linear workflows that involve moving data or triggering actions between well-known SaaS tools. It's got a low learning curve; its interface is friendly and you can set up a basic "Zap" (a trigger that leads to an action) in a few minutes. Its greatest strengths are its large app integration library and ease of use. An example use case is a new form submission. When the form is received (trigger), Zapier automatically creates a new contact in your CRM and sends you a Slack alert (actions). Zapier is great for those one-to-one handoffs and straightforward automations that glue your everyday apps together.

- **Make.** A visual workflow builder that allows more complex logic and branching. It's best for designing multi-step processes or automation scenarios with conditional paths. While it has a modest learning curve, the visual interface is quite intuitive. Taking full advantage, however, might require some trial-and-error and understanding of how to format data between steps. Its strengths include flexible logic (like if/then branches, loops) and powerful in-editor tools for transforming data (e.g., text parsing, scheduling delays). An example use case is an onboarding sequence where a single trigger (say, a client e-signed a contract) leads to multiple downstream actions, like updating a project board, sending a personalized welcome email, generating a project plan draft, and alerting the team in Slack or Teams. With Make, you can build that in one unified flow with clear logic between each step.

- **n8n.** An open-source automation tool which you can self-host, giving you full control. It's great for developers or tech-savvy consultants who need maximum flexibility and don't mind a slightly higher learning curve, though learning opportunities are plentiful (including company-hosted tutorials, third-party YouTube videos, and training courses from Udemy). While n8n requires a bit more technical understanding (and possibly hosting setup) compared to Zapier or Make, it rewards you with extreme customization capabilities, and creating simple workflows are a snap using its drag-and-drop interface and intuitive integration process. A key strength is its ability to accommodate custom functional nodes or to write custom JavaScript functions embedded in out-of-the-box nodes for added flexibility. It also allows connection to an API of virtually any tool (even if it's not officially supported), and enables you to keep your data on your own server for privacy if you choose the self-hosted option. One example use case might involve parsing meeting notes and extracting action items using an AI service. For example, after a Zoom call, n8n could take the transcript, send it to OpenAI's API for summary, then push the distilled action items into a separate spreadsheet or database app. This kind of custom integration is where n8n really shines.

Other helpful tools and services to consider (not necessarily AI-driven) include:

- **Calendly (or other scheduling apps).** Automate meeting scheduling by letting clients pick times from your availability, then integrate it. For example, when a meeting is booked, Calendly will automatically send a confirmation email and populate your calendar. You could extend this functionality by using a workflow tool to generate a meeting prep task in advance of the appointed meeting time.

- **Notion + AI.** Use Notion's databases and AI features to autogenerate project dashboard updates or summarize meeting notes within your workspace. For example, you could have a Notion page that, when updated with raw notes, uses an AI block to produce a summary or action list.

- **Airtable (with automations).** Airtable's spreadsheet-database hybrid is great for tracking deliverables. Its built-in automation can sync data with clients or send notifications. For instance, if you use Airtable to track project status, you can configure an automation that emails the client whenever a milestone status changes to "Completed."

- **PandaDoc or Qwilr.** Document automation tools that let you create proposal and contract templates that populate with deal info and can be sent for e-signature with minimal effort. When integrated with Zapier or others, you can have a new proposal go out as soon as a lead completes an intake form, for example.

- **CRM workflows.** Many CRM systems (Salesforce, HubSpot, etc.) have their own automation engines. You can often automate CRM tasks *within* those platforms—like sending a drip email sequence after someone is marked as a prospect, or creating a follow-up task three days after a meeting. Leverage these built-in features so you're not duplicating effort externally.

To be sure, automation shines brightest in the gaps *between* people and tools, when moving information from one system to another, or triggering next steps without waiting for someone to remember to do it. Whenever you notice "I export X from Tool A and import it into Tool B" or "after I do Y, I always send a reminder email," that's likely a handoff worth automating. Also, remember that there's no one-size-fits-all solution; choose tools that align with your consulting style and needs. If you prefer simple and quick, start with Zapier; if you love visual mapping, Make might be a better fit for you. If you're technically inclined and want greater flexibility, go with n8n. The key is selecting platforms you feel comfortable with and that integrate with the apps you already use (being mindful of cost considerations, of course). The best automation tool is the one you'll use consistently. Don't get caught up in the hype or what everyone else is using; stay authentic to what fits your workflow.

Matching Tools to Tasks

With so many tools available, how do you know which to use for a given job? While there's overlap, certain tools naturally lend themselves to certain tasks. To help you decide, here's a quick pairing guide matching common consulting tasks to appropriate tools:

Task	Tools	Rationale
Client intake & onboarding	Zapier + PandaDoc	Easily send intake forms and auto-generate contracts for new leads, providing a seamless welcome.
Proposal creation	Make + Google Docs	Fill proposal templates based on form responses or database info, allowing complex logic (e.g., different content blocks) if needed.

Task	Tools	Rationale
Calendar sync & follow-ups	Calendly + Zapier	One-click scheduling for clients, then Zapier triggers follow-up emails or tasks after an event is booked or completed.
Transcription & summarization	n8n + OpenAI API	Automatically transcribe meetings (via Zoom API or Otter.ai) and summarize key points with GPT, then share notes.
Task tracking & status updates	Notion + Make	When a new document or item is added (e.g., a client uploads a file), update project status in Notion and notify the team, using Make for the integration logic.

These pairings are not exclusive, but they're a great starting point. Often you'll combine multiple tools—for instance, an intake process might use Typeform (for a questionnaire) to trigger a Zapier "Zap" that populates a Google Doc proposal that is sent to PandaDoc for electronic signature. Mix and match based on where your source data is and where it needs to go. The important part is understanding *why* a tool fits a task. Calendly, for example, specializes in scheduling (so use it for anything calendar-related), whereas Zapier's strength is connecting different apps after a simple trigger (great for sending data from one app to another on an event). If you choose the tool whose sweet spot matches the task, you'll get better results with less effort.

Common Consulting Automations

Let's look at some real-world use cases of automation in a consulting workflow. These examples illustrate how you might streamline various parts of your practice:

- **Proposal generation.** *Automate your sales proposal creation.* For example, a prospective client fills out a web form or questionnaire (providing their needs, budget, etc.). That submission automatically populates a proposal template with the client's details and specific service scope. The system then generates a polished PDF and even drafts a personalized email ready to send to the client. Instead of spending an hour drafting a proposal from scratch, you only need to review and approve the auto-generated version. The client gets a rapid response with a full proposal perhaps within hours of their request—a huge differentiator when speed matters.

- **Discovery call follow-up**. *Never forget to send meeting notes again.* Imagine your Slack or Teams client call recording in Fathom is automatically transcribed, summarized with key points and action items, and automatically sent to all meeting attendees *seconds* after the meeting ends. A Zapier workflow then logs a copy to your CRM or notes system for next steps. This means every client conversation has a reliable follow-up, and you've provided value by distilling the discussion without spending your evening transcribing and organizing notes. Consultants using such AI-driven assistants have been able to save several hours per week on documentation while ensuring clients feel heard and looked after.

- **Client check-in sequences.** *Maintain ongoing relationships with light-touch automation.* A common practice for coaches and consultants is to periodically check in with past or current clients to see how things are going. Automation can handle the cadence and initial outreach. For instance, you might set up a sequence where thirty days after a project concludes (or between major project milestones), the client gets a friendly, personalized email that says, "Hi [Name], it's been a month since [project/meeting]. I was thinking about how [their goal] is progressing. How are things going? Let me know if you need any support." You can write these notes in your tone, and the system will populate any specifics (like project name or goal) and send it out on schedule. If the client responds, you take over personally. If not, the automation can even remind you to follow up via phone after a certain number of touchpoints. This keeps the relationship warm and shows clients you haven't forgotten them. Many consultants find that such scheduled check-ins lead to repeat engagements or referrals, all initiated by an automated nudge that takes no additional time on your part.

- **KPI dashboards and reporting.** *Deliver insights to clients automatically.* Suppose your services include weekly or monthly updates on key metrics (marketing performance, financial metrics, etc.). Gathering that data and formatting a report can be automated. You could develop a system that automatically pulls data from various sources (e.g., Google Analytics, a client's CRM, a financial spreadsheet, etc.) and compiles an update email or dashboard every Friday. It could generate charts or a summary of week-over-week changes, and then email it to the client team, providing them with timely insights without scheduling an extra meeting. This also forces you to have a consistent pulse on client data, and the client enjoys quicker turnaround and more frequent visibility into progress. Of course, you'd review the auto-report first to ensure accuracy, but over time it becomes sufficiently refined to run on its own. This not only saves time, but also improves the client experience by providing faster, more reliable service delivery.

- **Project status updates.** *Keep internal and client teams aligned effortlessly.* Consider a scenario where various project documents are being produced (e.g., a strategy document, a research report, a draft roadmap, etc.) and stored in a shared folder. Normally, someone might have to manually update a status tracker or send an email stating that "Deliverable X is ready for review." With automation, each time a new file is added or updated in the project's folder (the trigger), an update can be made to your status board (e.g., an Airtable or Asana board) marking that deliverable as done. Simultaneously, a notification can go out to the client (perhaps via email or Slack) saying, "The Q3 Strategy Draft has been uploaded for your review." This keeps everyone properly informed in real time. No more wondering whether a client has been informed of a key milestone or completion deadline being met, because the system has already informed them! This reduces communication lags and ensures transparency.

These are just a handful of examples. In practice, anything repetitive or scheduled in your workflow is a candidate for automation. Be creative; think about the tasks that steal minutes from every day, or the processes that, if improved, would noticeably enhance your service. You might surprise yourself with how many parts of your consulting routine can be streamlined.

Automation as a Differentiator

Let's examine the case of a hypothetical process improvement engagement wherein you've set up an elaborate workflow using Make that illustrates the power of automation. Rather than manually writing up each Standard Operating Procedure (SOP) after manually taking notes during staff discovery interviews, you could record the interviews and feed the audio into a transcription and formatting automation tool. The system could then transcribe each interview, extract key steps and decisions (using an AI text analysis), and auto-generate a draft SOP document complete with a consistent template and formatting. What used to take many hours could be completed in a fraction of the time, with only light editing needed on the drafts. Clients will be astounded at how fast and thoroughly you can turn their conversations into formal documentation. Accordingly, automation not only saves time, it also becomes part of your value proposition. By showcasing innovative automation in your service delivery, you differentiate yourself from competitors and can even turn your internal automations into client-facing offerings.

Automating the Client Journey

To really see the impact of automation, it helps to view it holistically across an entire client engagement. Let's walk through a typical consulting engagement and highlight opportunities to inject automation at each stage. By mapping the client journey from initial contact to project close and beyond, you can design a seamless experience where automation supports both you and your client every step of the way.

Pre-engagement (lead to proposal). This is where first impressions are made. If a potential client reaches out or fills out a form on your website, automation can ensure an immediate and helpful response. For example:

- **Lead capture to automated response.** The moment a lead submits an inquiry or downloads a PDF from your site, an automation sends a thank-you email acknowledging their interest and perhaps provides a link to schedule an introductory call. This could be done via a CRM or email marketing tool integrated with your site. The prospective client gets near-instant feedback (even if it's 2:00 am on a Saturday), demonstrating your responsiveness well before work begins.

- **Intake form to CRM update and prep.** Before an intro call, you might have the lead fill out a short questionnaire. Once they do, automation can push those details into your CRM and even create a one-page briefing document for you. For instance, the answers could be merged into a Notion or Word template so you have a nicely formatted lead sheet summarizing their needs when you hop on the call. This saves you the prep work of compiling their info, and nothing they submitted will be overlooked when you talk.

Engagement kickoff (onboarding). After the prospect becomes a client, kickoff is about turning that signed deal into a running project. There are many minor tasks here that automation can handle to ensure a smooth start.

- **Contract signed to kickoff checklist to client.** When the client signs the engagement contract (perhaps via PandaDoc or DocuSign), you can have an automation trigger that sends them a welcome packet or checklist. For example, an email might go out immediately thanking them for coming on board, outlining next steps, and perhaps requesting any initial data or scheduling the kickoff meeting. This ensures the client isn't left waiting and wondering what happens next; they get an immediate, professional briefing.

- **Internal checklist to project workspace setup.** Likewise, internally you likely have a list of things to do when a project kicks off (e.g., assign team members, create a project folder or workspace, set up a Slack channel, etc.). Automate as much of this as possible. For instance, a signed contract could trigger creation of a project in your project management tool (like Asana or Trello) using a template for that type of project. Or using Make, a new Notion project page could be generated pre-filled with the client's name, project scope, key dates, etc. All the housekeeping happens behind the scenes, so your team can hit the ground running without missing any setup steps.

Delivery phase (execution and ongoing work). During the core of the project, the focus is on delivering results, but there are plenty of recurring tasks that glue the process together. Consider the following use cases:

- **Before meetings: Automated prep briefs.** Schedule systems can trigger helpful meeting preparation. For example, the night before a client workshop, an automation could compile relevant materials (e.g., the latest status report, last meeting notes, any open issues) into an email or document for you and the team. It might pull the latest data from a dashboard and list the key points to cover. You wind up with a meeting brief waiting in your inbox in the morning with zero effort, ensuring you walk in prepared without any last-minute cramming.

- **After meetings: Instant notes and next steps.** We discussed this in the follow-up example earlier—capturing notes or action items via automation. Whether it's through a transcription service in Zoom or Fathom that adds an AI summary, or even a structured form you quickly complete that triggers updates to a distribution list, automating meeting follow-up communications means clients get near-real-time documentation. By the time a client drives back to their office after a workshop, they might find a meeting summary and key action items as next steps in their inbox. That kind of responsiveness, enabled by automation, is often beyond what even larger firms provide; it really stands out, and many products handle all of this natively.

- **Recurring updates: Push instead of pull.** If you have weekly check-in calls or status emails, automate the creation of those updates. Every Monday, a simple tool could pull the latest task status, budget spent, or KPI values and draft a status email for you using a custom GPT. You review and add any commentary or context, then send. The heavy lifting of gathering the data is done automatically. This way, both you and the client always have up-to-date info with minimal lag.

Post-engagement (project close and follow-up). When the formal project work is done, automation helps you leave a lasting positive impression and can sow the seeds for future business. Here are some examples designed to improve this phase of your consulting work:

- **Feedback & closure at project completion.** Once you deliver the final report or deliverable, kick off a wrap-up sequence. For instance, initiate a "project closing" sequence that emails final documents via a link while triggering an automated feedback survey to the client. Their responses can then be forwarded to a testimonial file or suggestion box. (Many of today's LLMs are quite capable of separating praise from criticism, and routing accordingly using workflow automation platforms like Zapier or n8n.) Also, an automation can mark the project "closed" in all relevant systems, archive files to the right folders, and even generate a summary of the project outcomes for your internal knowledge base.

- **Scheduled check-ins for relationship nurturing.** Don't just leave the client in the dark after the project. As mentioned earlier, it's a good practice to schedule a timely (thirty to sixty days post closure) follow-up email automatically. For example, a month after project closes, the client might receive a message stating, "It's been a month since we wrapped up; just wanted to check in to see how things are going. Did the recommendations prove helpful? I'm here if you have any questions or could use a sounding board." This message can be personal in tone but sent without you having to remember the date; your CRM or email tool does it for you. Clients often appreciate that you're thinking of them beyond just the paid engagement, and it keeps the door open for next steps. If nothing else, it reinforces that relationship-first mentality so characteristic of exceptional advisors.

At each of these stages, the consultant's touch is still there guiding the content and strategy, but automation is acting like an unseen assistant, handling the timing and execution of tasks. The result is a smoother journey for the client with timely responses, organized processes, and fewer gaps—and less administrative burden for you.

Designing for Automation

When you're ready to build an automation, plan before you platform. Spend some serious time designing the workflow before jumping into the tool to create it. A common pitfall is to automate a process that's poorly defined or inefficient to begin with. As the late, great process improvement expert Michael Hammer put it, "Automating a mess yields an automated mess." First streamline and clarify the process, then layer automation on top. Taking this deliberate approach ensures that automation amplifies a *good* system rather than enshrining a bad one. Here are a few tips for effective automation design:

- **Map the process end-to-end.** Start with a simple flowchart or checklist of the manual process as it exists (or should exist). Identify each step, who is responsible, what tools are used, and what information passes through. This mapping helps you see the big picture and all the decision points. Often, drawing it out can reveal unnecessary steps or clarify where an automation trigger should happen. For example, sketch out the steps from receiving a client inquiry to having a scheduled intro call; seeing it visually makes it easier to decide where to introduce automation.

- **Identify friction points.** As you map the process, mark the steps that are slow, inconsistent, or prone to error. Ask, "What causes delays or headaches here?" Maybe approvals sit in someone's inbox too long, or data gets entered wrong from form to spreadsheet. These pain points are where automation can add value by either eliminating that step or performing it more reliably. For instance, if you often forget to send a follow-up, that's a friction point and a solid candidate for an automated reminder.

- **Choose clear trigger points.** Every automation needs a trigger—an event that starts the workflow. Decide *exactly* when the automation should run. Is it time-based (e.g., every Friday at 5:00 pm)? Is it event-based (e.g., a form submitted, an email received, a task marked complete)? Pinpointing the trigger is crucial. A good trigger is specific and observable by your tools. *"When a new row is added*

114

to the spreadsheet" or *"when the calendar event ends"* are concrete triggers. Make sure the trigger chosen aligns with the desired outcome. For example, if you want an action right after a meeting, the meeting's end time can serve as a trigger via the calendar app's integration.

- **Define success criteria.** How will you know the automation worked correctly? Define the expected result, so you can verify it. For example, *"'Success' means the client receives the summary email and the action items are logged in the project board."* By knowing this, you can test the automation and check those outputs. Essentially, treat it like you would a deliverable, i.e., by ensuring it has acceptance criteria. If something is off (say the email format is wrong or the data is misaligned), you can adjust the design. Having an unambiguous definition of "done" also prevents scope creep in building the automation, because you'll know exactly what it needs to accomplish.

- **Incorporate error handling or notifications.** Even well-built automations can fail. An app's API might be down, a data format might be unexpected, or a step might quietly do nothing because a field was blank. Design with failure in mind. This could mean setting up an alert. For example, if an automated email is stuck in the 'outbox' for more than five minutes, notify me via Slack. Or add a step that logs the automation's run and result to a spreadsheet or database for later review. Many platforms let you create conditional "if this error occurs, do x" paths. At a minimum, periodically check the logs of your automations. By monitoring, you'll catch issues early. Think of it as managing a junior assistant; you want to know if they've run into a problem as soon as possible. For instance, if an automated report generation fails because data was missing, have the system flag it so you can manually intervene. A silent failure can be worse than no automation, because you (and the client) might be assuming something happened when it didn't.

- **Pilot on your own or with a test case.** Before fully rolling out an automation (especially client-facing ones), test it in a low-risk environment. Use your own email address as the client in a trial run or run the automation with a small internal project first. This helps ensure everything works as intended. It's much better to catch a glitch when you're acting as the client than to have a real client annoyed by an irrelevant or incomprehensible email. During testing, deliberately try to break the workflow. For example, see what happens if a form field is left blank or if two triggers happen simultaneously; this helps harden the process. Only once you've seen it perform reliably should you promote it to production.

When designed well, every automation should feel like a behind-the-scenes team member who's quiet, reliable, and invisible until needed. It does its job on cue and then stays out of the way. Achieving that level of operation requires thoughtful design upfront, but it pays off when your automations run week after week without issue. Remember, the aim is to *enhance* your workflow, not complicate it. So start with a clear, well-thought-out process, and your automation will fit right in as a helpful assistant.

In fact, you might find it useful to give your key automations a "codename" or nickname, almost as if they were actual team members. For example, "Client Onboarder" or "Weekly Wrap-up Bot." This isn't just cute—it helps you think of them as parts of your operation that need care and feeding. Just like you'd onboard a team member, train them, and check their work initially, do the same for your automations. Soon, they'll fade into the background, quietly doing their jobs just like the best support staff.

Governance, Ethics, and Human Oversight

Automation adds speed, which can add risk if left unchecked. When you're moving fast on autopilot, it's easier to crash into a mistake if you're not steering. Responsible use of automation is critical, both for your sake and your clients'. Here are some best practices to ensure your helpful automations don't inadvertently create problems:

- **Test thoroughly before going live.** As mentioned, always pilot your automations in a safe environment. Run multiple scenarios. If your workflow sends an email or file, double-check the content and recipients in a test. Does the right data appear? Is the tone correct? Is it going to the intended people? Consider edge cases (empty fields, unusual inputs) and see how it behaves. This upfront QA process will catch most issues. Essentially, you're doing the same kind of testing you'd do for any important process or for delegating to a new assistant—making sure they know what to do in all situations.

- **Start with human oversight for critical steps.** In the beginning, or for very sensitive tasks, keep a human in the loop. For example, maybe an automation prepares a draft email or report, but you review it before it goes out. Or an automation schedules a tweet but you have to approve it before it gets posted. This hybrid approach ensures nothing catastrophic happens without your knowledge. Over time, as you gain trust in the automation, you can loosen the reins. But never fully "set and forget" critical communications or decisions. Maintain manual checkpoints for anything that could have serious repercussions if it misfired (like a wrong number in a financial report or a client email that could be misinterpreted).

- **Log automated interactions.** Keep records of what your automations do. Most platforms have logging, but you might also create your own log (e.g., a Google Sheet where each row is "Automation X ran at [time], did [result]"). This is important for accountability and debugging. If a client says, "We never got the report last week," you should be able to confirm if it was sent and when. Logging also helps you demonstrate transparency; you could even share certain logs with clients to show the consistency of your process. It provides a defensible audit trail. In regulated industries or larger consulting engagements, this can be crucial. For instance, if you automated data analysis steps, logging what was done and when can help in validating your findings or tracing any discrepancies later.

- **Mind data security and privacy.** When automating, you might be piping data through third-party services. Be very conscious of client confidentiality and any regulations that apply. Only use reputable tools that have proper security measures. Ensure data is encrypted if needed. If you're using an AI service to summarize a client's document, for example, consider whether that data is allowed to be sent to that service. Some firms choose self-hosted or on-premise tools for sensitive information for this reason. Always be transparent with clients about what's automated and where their data might be going. Most clients will understand (or even appreciate) that you use certain tools, especially if it benefits them, but they should never be caught off guard. A simple disclosure like, "We use a secure third-party service to transcribe our meetings" in an agreement can suffice.

- **Watch out for bias or errors in AI outputs.** If your automation involves AI (say, generating text or analyzing data), keep in mind that AI is not infallible. It can introduce subtle errors or biases. For example, a language model might slightly mischaracterize a discussion or use an off tone in an email draft. You, as the human expert, must supervise these outputs. Regularly review the content AI produces. Make sure it aligns with your voice and facts. AI can also reflect biases present in training data, so be cautious in contexts like analyzing people-related data or making recommendations; ensure the results are fair and sensible. Essentially, don't outsource your critical thinking. Use AI to assist, but apply your consultant's judgment to all final outputs.

- **Establish an update and maintenance routine.** Just like software, automations require maintenance. Apps change their APIs, your processes evolve, or business rules shift over time. A rule you set up today might not hold in six months. Make it a habit to review your automations periodically, like once per quarter. Check that they're still doing what they should do. Update any references (e.g., if you switched project management tools, update the integration). Also, whenever you change a business process, ask yourself whether you need to modify the automation. Treat automations as part of your operations that need consistent care. Document them, too; keep a simple list of active automations and what they do. If you take a vacation or hand off to someone, it's useful to know what's running in the background.

- **Be transparent and ethical.** Let clients know, at least in broad terms, the automations that touch them. For example, you might mention in your kickoff, "We have systems that will send you automatic updates every Monday" or "Our scheduling and follow-up are automated to ensure you always get timely information." Framing it as a benefit is easy because it truly is one. Transparency builds trust; clients won't be puzzled when they get an email that obviously wasn't typed by you at 1:00 am, and they'll appreciate the consistency. Ethically, avoid automating anything that would mislead. Don't have a bot tweeting personal-sounding insights as if it were you, or an AI writing personal client messages that pretend to be hand-written. It's a fine line; using templates and automation is okay if it's essentially the same as what you'd do manually. Just be cautious not to cross into deception. Keep the human-authored vs. machine-authored distinction clear, especially in client deliverables.

Finally, treat automation with the same care as delegation. If you'd inform a human assistant of schedule changes or special cases, you must do the same for your automated "assistant." And always have a way to quickly disable or edit client-facing automations if something changes. Consider implementing checklists. For example, at project close, check to ensure any scheduled automated emails are either appropriate or canceled to avoid any surprises.

In summary, governance and oversight are about keeping you in control of your automated processes. With sensible precautions, you can enjoy the efficiency gains without undue risk. Think of automation as driving a faster car—you still need good brakes, seatbelts, and a dashboard to monitor the engine. Build those in, and you can ride in confidence.

Scaling Automation Gradually

After reading this, you might be excited to automate *everything*! That enthusiasm is great—but remember the principle of incremental improvement; don't try to automate all at once. Incremental adoption is the sustainable path. Start small, learn, adjust, and then expand. Experts recommend taking a phased approach where you tackle one area at a time, demonstrate value, then build on it. This avoids overwhelming yourself (or your team) and ensures each automation has real impact. A practical way to phase your rollout is to start with one automation in each of three areas—one internal, one client-facing, and one follow-up. For example:

- **Internal task automation.** Pick a purely internal process that bugs you—for example, preparing your weekly team meeting agenda or tracking your expenses. Automate that first. This is low-risk (no client sees it) and gives you a quick win. Perhaps you create a Zap that collects all tasks marked "done" every Friday afternoon for that week and emails you a summary for your Monday morning meeting, providing an instant agenda draft.

- **Client-facing step automation.** Choose one point in the client experience to enhance. Scheduling is a common one—if you haven't already automated calendar bookings, that's a good start. Or maybe the meeting follow-up notes as described earlier. Implement it for one client or one type of meeting first. Monitor the client's reaction. Are they pleased with the speedy notes? Did the scheduling link reduce back-and-forth emails? Use that feedback to refine your approach.

- **Follow-up or nurture automation.** Put in place a simple follow-up sequence, like the thirty-day check-in or a monthly newsletter if you have multiple clients. Again, start with a subset—perhaps just for clients from the past quarter. See how it goes, ensure it's sending appropriate content, then expand to more contacts. This not only tests the automation, it tests your comfort. You might realize, for instance, that a biweekly cadence was too much and adjust to bi-monthly.

By starting with these three, you cover different dimensions (your internal efficiency, client service during a project, and client relationship after a project). They act as pilot projects for your broader automation journey. As you gain confidence and see the benefits, you can deepen each area and add more processes to automate.

Your Automation Starter Kit

To wrap up the "getting started" part, here's a quick checklist to guide your initial automation projects:

- **Identify a few high-friction, repeatable tasks.** Use your inventory and ROI expectations to choose the best candidates. Aim for tasks that annoy you *and* happen often (daily/weekly).

- **Choose a platform that fits your needs and skills.** Based on the task, decide which tool is best (Zapier, Make, n8n, etc., or an in-app automation feature). Stick to one platform for your first project to keep things simple.

- **Map and prototype the workflow manually.** Before automating, walk through the process manually one last time, or simulate it. Make sure you know each step. For example, manually send that follow-up email once and note what info you included. This acts as a blueprint for the automation.

- **Build the automation and test it in a sandbox.** Set up the workflow in your chosen tool, using test data. Run it and see if it does what's expected. Iterate until it's smooth.

- **Launch with a small scope.** Turn on the automation for a small case—maybe one client, one project, or your own internal use. Monitor it closely the first few runs.

- **Monitor results and gather feedback.** Check the logs and make sure the actions occurred correctly. If it's client-facing, subtly ask the client if they found the update helpful (or watch for their reaction). Pay attention to any glitches.

- **Iterate and expand.** Fix any issues, adjust timing or messaging if needed, and once it's working well, consider rolling it out more broadly (to all similar projects or to more clients). Then move on to your next automation project, applying what you learned.

By following this checklist, you develop your "automation muscle" gradually. Each success builds momentum and confidence. Importantly, this approach also builds organizational buy-in if you work with a team; you can show colleagues the clear benefit of the automation you implemented, which will get them on board with the next one. It's much easier to scale automation when everyone has seen it working well on a small scale first.

Remember, the journey to an automation-augmented practice is a marathon, not a sprint. You're establishing a new way of working, and that takes iterative improvement and sensitivity to change resistance. As you progress, what felt advanced will become routine, and you'll spot many more opportunities to automate further. Just keep the pace that allows thoughtful implementation.

Free Time is a Strategic Asset

Automation isn't about doing less work—it's about making space to do more of what truly matters. By delegating robotic tasks to actual robots (or code), you reclaim time and mental energy for the uniquely human aspects of consulting. Imagine if you gained an extra ten hours a week because your "digital assistants" handled the grunt work. How would you use that time? Perhaps diving deeper into research and discovery for clients, or honing insights that lead to breakthrough strategies? Maybe spending more face-to-face (or Zoom-to-Zoom) time with clients, asking questions and listening to strengthen those relationships? Or simply having more bandwidth to be creative and think outside the box for solutions?

Freeing time allows you to elevate your role from automaton to true strategic partner. Instead of racing to get a report done at midnight, you can sleep on it and add sharper analysis the next morning. Instead of rushing through a client call because you have three hours of admin tasks to do afterwards, you can be fully present and attentive, knowing much of that admin will handle itself. The compounded effect over an engagement is significant:

- **Deeper discovery.** You can invest more time in understanding the client's context and problems, which leads to more insightful recommendations. Automation gives you the breathing room to *really listen* and probe, rather than thinking about the backlog of tasks.

- **Sharper insights.** With AI and automation gathering and processing data, you get information faster and can spend time synthesizing it. This means your analyses and strategies can be more data-driven and well-thought-out, not superficial. You're leveraging tools to crunch numbers or draft background research, then applying your brain to interpret what it means for the client.

- **More present facilitation.** Whether it's a client workshop or an executive meeting, you can focus on the dynamics in the room, facilitate effectively, and read the nuances—rather than taking notes or setting up technology. You've outsourced those latter jobs to automation (e.g., a meeting recording tool), so you can concentrate on guiding the discussion and coaching the client.

- **Thoughtful follow-up.** Instead of generic or delayed follow-ups, you can craft truly thoughtful messages and next steps. You might use the extra time to personalize a thank you note or follow up on a particular comment a client made. Ironically, by letting automation send the basic template stuff, it frees you to add a human touch that is far more meaningful. Clients will notice the care in your follow-ups.

At a broader level, by streamlining operations, you also scale your impact. You can take on more projects or devote time to developing new offerings. Perhaps you use the hours saved to develop assets (e.g., articles, frameworks, workbooks, etc.) that further differentiate you. Or you can simply focus on your own wellbeing to prevent burnout and stay creative and enthusiastic in your work. All of these outcomes lead to a more sustainable and growing consulting practice.

It's telling that major consulting firms are heavily Investing In AI and automation. According to IBM's *AI Tools for Consulting* resource, "AI and automation [enables] consultants [to] significantly reduce the time needed to gather and analyze data, leading to more productive client interactions and quicker project turnaround times." That isn't a dire warning; it's an exciting opportunity. It means half of what we busy ourselves with today might be handled by our tools, allowing consultants to focus on higher-order work like change management, innovation, and human-centered advising. The very nature of the consulting job will shift toward the aspects that AI cannot do, like building trust, making judgment calls in ambiguous situations, and empathizing with client needs. By embracing automation now, you're positioning yourself for that future, where your role is even more about brainpower and heart, and less about paperwork.

So, start small, but start now. Pick one internal process and automate it. Map it, build it, test it, and experience the excitement when it runs on its own. Then reinvest that time wisely. Double down on client engagement or thinking deeply about a client's problem. *Then* go to the next automation. Step by step, you'll construct a tech-augmented practice that operates smoothly, delights clients, and amplifies your strengths rather than dulling them.

And there's a philosophical underpinning here. The most successful consultants in the age of AI will be those who integrate technology in a human-centric way, using it to augment their authenticity and creativity, not replace it. Automation is your ally in designing a consulting business that aligns with your values (i.e., you decide what to automate based on what you consider low-value work) and enhances your client relationships by ensuring reliability and more personal attention where it counts. In the end, there is no one-size-fits-all blueprint for this. You have the freedom to craft automation solutions that fit your practice, your personality, and your clients.

As you experiment and iterate, you're innovating your own way of working. In doing so, you fulfill one of the central ideas of this book: that AI and automation should empower you to chart your own path as a consultant, delivering value in a way only you can, supported by the best tools of the day. Embrace automation as a creative tool, and there will truly be no limit to the impact you can have.

◆

CHAPTER SEVEN

AI-Powered Meeting and Note Taking Tools

C onsultants don't typically aspire to take notes, but they do need an accurate record of the many conversations that inform their findings and recommendations. Such a record enables them to make sense of conversations, connect dots, and guide decisions. AI meeting tools, ranging from Otter.ai to Fireflies to Fathom, have evolved from basic transcription services into powerful collaboration partners. Used well, these tools free consultants from tedious notetaking and allow them to listen more actively, reflect more deeply, and respond more strategically. In other words, AI notetakers handle the minutiae so you can focus on higher-value work. This chapter explores how AI-powered meeting assistants help consultants reclaim cognitive bandwidth, improve follow-through, and enhance client trust.

Why AI Note-Taking Matters for Consultants

The modern consultant often juggles multiple projects, stakeholders, and contexts. Critical insights can slip through the cracks when you're scrambling to type everything said. AI meeting tools help consultants to:

- **Capture accurate transcripts without distraction.** Instead of splitting attention between listening and writing, you get a complete record of the conversation automatically.

- **Summarize key themes, decisions, and action items.** Advanced AI can condense an hour-long discussion into the salient points, saving time on recap.

- **Generate client-ready recaps with minimal effort.** Many tools now draft follow-up emails or summary reports for you.

- **Preserve institutional knowledge across projects.** Transcripts and highlights are stored, creating a searchable memory bank of client meetings.

- **Stay focused on the moment instead of your keyboard.** With AI handling notes, you can maintain eye contact and actively listen, improving the quality of conversation.

I've used Fathom to automatically summarize meetings and distribute meeting highlights and action items to all participants immediately after the meeting ended (with the meeting owner's permission, of course). Instead of frantically scribbling notes, I was able to be fully present with the client. Before subsequent meetings, I'd quickly review Fathom's highlights to recall exactly what was discussed and which action items needed attention. Clients seemed to love how organized I was from session to session, because nothing fell through the cracks. By outsourcing notetaking to AI, I've been able to deepen trust and maintain continuity with my clients from meeting to meeting.

The Bandwidth Equation

Taking notes consumes executive function. Psychologists note that our working memory and attention are limited. For example, doing two things at once (i.e., listening and writing) adds cognitive load. When consultants offload the note-taking task to AI, they free up mental bandwidth for reading tone, observing body language, and synthesizing meaning in real time. Retention of information *and* presence in your discussions are both dramatically improved.

A Quick Comparison of Key Tools

Here's a snapshot of some popular AI meeting assistant tools and their strengths as of this writing. All of these platforms can record meetings, transcribe dialogue, and generate summaries, but their focus, integrations, and AI capabilities vary.

- **Otter.ai.** A well-known note-taking app that now goes beyond transcription. Otter's AI Meeting Agent can auto-join Zoom, Microsoft Teams, and Google Meet calls to provide live transcripts and even real-time summaries and action item tracking during the meeting. Uniquely, Otter introduced a cross-meeting knowledge base—you can ask its chatbot questions across *all* your past meetings and get answers drawing on those transcripts. It offers both free and paid plans, with the paid plans offering more hours. Otter.ai is great for real-time meeting insights and a searchable archive of conversations. Note that Otter only supports English, French, and Spanish (as of this writing), so it's not yet a multilingual solution. Also, its focus is on meeting content; it lacks built-in CRM integrations or sales intelligence features.

- **Fireflies.ai.** A robust AI notetaker favored by many teams for its breadth of integrations and analytics. Fireflies can automatically join meetings across most popular platforms including Zoom, Teams, Meet, Webex, and Slack, and transcribe in more than sixty languages, making it very friendly for global teams. It delivers real-time or near-real-time transcripts (up to ~95 percent accuracy under good conditions) and generates concise summaries that highlight action items and decisions. Fireflies stands out for its CRM integration; it can log notes and tasks directly into systems like Salesforce or HubSpot, which is a boon for sales and account management use cases. It also provides conversation intelligence metrics like speaker talk time and keywords mentioned for further analysis. It's great for multi-language support and workflow integration (especially if you want meeting data synced to other apps), but beware that while the interface has a lot of features, it can be overwhelming for new users. Some advanced analytics require a premium plan, though Fireflies does have a free tier for basic use.

- **Fathom.** An AI meeting assistant originally built as a Zoom add-on, now compatible with Google Meet and Teams as well. Fathom is notable for its generous free plan, which includes unlimited recordings and transcripts for individual users. During a call, Fathom lets you mark key moments with a click (e.g., highlight an important point or task) and later automatically produces a summary of those highlights. It even offers a variety of summary formats and can generate follow-up emails and action item lists from the call. In fact, Fathom provides multiple AI-generated summary styles tailored to different meeting types (e.g., sales call, product feedback, etc.). It's best for individuals or small teams on a budget, and those who want quick highlights, summaries, and action items for every call. Fathom may not have some of the enterprise features of others (though it is SOC 2 and GDPR compliant). Also, when using Fathom via the Zoom app, other participants will see a banner that Fathom is "in" the meeting, so you'll want to be transparent and ensure that's okay.

- **Sembly AI.** A newer entrant focused on turning meeting data into deeper insights. Sembly records and transcribes meetings across major platforms such as Zoom, Teams, Google Meet, and Webex, and produces comprehensive meeting minutes by default. What sets it apart is the advanced AI analysis layer; Sembly can automatically identify and track action items, issues, decisions, and key topics from discussions. In its latest version, Sembly introduced a multi-meeting AI chat that lets you query across multiple meeting transcripts at once (similar to Otter's knowledge base idea). It also offers AI Artifacts, which automatically draft documents like project plans or proposals based on your meeting content, and personalized post-meeting insights that recommend next steps for you based on the discussion. It's very good for enterprise teams or project managers who want analysis and outputs from their meetings in addition to standard notes. Be aware that Sembly's cutting-edge features come with a cost; the multi-meeting analysis and artifact generation are available on higher-tier (paid) plans. And while Sembly is enterprise-security focused, ensure your organization is comfortable with its data policies if you handle sensitive info.

- **Supernormal.** A rising star that brands itself as an "AI-powered meeting" platform rather than just a note-taker, Supernormal works across Google Meet, Teams, and Zoom via a browser extension or bot, and automatically transcribes and summarizes meetings. Its unique twist is helping with meeting preparation and in-meeting guidance. For example, Supernormal can generate AI-suggested agendas for your recurring meetings (like weekly team syncs or one-on-one meetings) so you start with a plan. During the meeting, its AI (nicknamed "Norma") can answer your questions in real time and highlight key points as they happen. It will then compile the notes and action items, and can even automatically share the meeting summary with attendees. Supernormal also provides templates for note-taking and a centralized repository for all your notes, making them easily searchable. Supernormal work well for managers and consultants who want an all-in-one tool that not only captures notes but also helps run the meeting (agenda, real-time Q&A) and follow up. Note that, because it actively interacts in meetings (e.g., by posting summaries or tasks), be sure all participants are on board with that level of AI involvement. Its free plan has limits on the number of meetings per month, with paid plans for heavier use.

It's also notable that Microsoft Teams and Zoom have their own built-in AI assistance, but those are walled garden solutions; they work only within those platforms. Tools like Otter, Fireflies, Fathom, Sembly, and Supernormal can join virtually any meeting on your calendar regardless of platform, which is a big advantage for consultants who hop between clients' preferred meeting apps. Also, when evaluating any tool, look for enterprise-grade security (encryption, compliance certifications). Many leading vendors are now SOC 2 and GDPR compliant to meet corporate privacy standards.

Use Cases Across the Consulting Lifecycle

AI meeting tools shine across the entire consulting process, from business development to delivery. To illustrate this, let's explore a few high-impact use cases.

Discovery Calls. Early conversations with prospective or new clients are packed with information. An AI notetaker can:

- Capture the client's exact language describing their pain points (great for echoing their words in proposals or marketing).

- Generate a post-call summary that flags the client's key needs and questions.

- Track questions the client asked *you* to give insight into their priorities or concerns.

Workshops and Facilitation. In interactive sessions or group discussions, it's hard to document everything in real time. AI can:

- Highlight insights or decisions in real time with hotkey presses (so you mark important moments without breaking the flow).

- Produce a transcript of multi-threaded discussions, helping ensure no idea is lost even if conversations overlap.

- Provide a rapid recap after the session, reducing the usual admin load of compiling notes.

Coaching and Advisory. For ongoing advisory engagements or coaching relationships, consistency is key. AI tools let you:

- Pull up previous session insights in seconds, so you can quickly refresh your memory on what was discussed last month or last quarter.

- Detect patterns in a client's language or sentiment over time, revealing trends (e.g., the client becoming more optimistic, or a recurring anxiety that surfaces each meeting).

- Automate delivery of session summaries and even prep notes. For instance, after each session you send the client a neat summary, and before the next session you review an AI-generated brief of last session's takeaways.

Project Management. When leading a project or working with a team, meeting assistants become a second brain for the team's communications:

- Extract action items and owners from project syncs or stand-ups, so nothing falls through the cracks.

- Sync decisions and tasks to a central source of truth (e.g., push action items to a Trello board or a Notion page automatically).

- Keep everyone aligned by sharing summaries, especially useful for team members in different time zones who couldn't attend live.

From Transcript to Insight

A transcript is not a deliverable. *Insight* is. In consulting, a verbatim transcript of a meeting has limited value on its own; the value comes from distilling meaning and next steps from that raw information. Modern AI meeting tools go well beyond basic speech-to-text. Many now offer capabilities to bridge the gap from transcript to insight. For example:

- **Automatic topic clustering.** The AI groups parts of the conversation by themes (e.g., objectives, concerns, decisions), which helps you identify the larger categories discussed.

- **Highlight tagging in real time.** You can tag moments as they happen (e.g., #Decision, #Risk, #ActionItem) to easily filter them later. Some tools auto-tag important moments for you.

- **Smart summaries by topic or speaker.** Instead of one giant summary, the AI might break down notes by agenda item or by person, which is quite useful if different stakeholders care about different sections of the discussion.

- **Sentiment and tone analysis.** Advanced meeting intelligence can analyze how things were said, identifying enthusiasm, hesitation, interruptions, or even detecting if a speaker's tone was optimistic or concerned.

- **CRM or app integrations.** Some meeting assistants can automatically update other systems. For example, they might log a call summary to your CRM contact record, or create tasks in your project management app when action items are mentioned.

Be sure to customize your summaries; don't just accept the AI's first draft blindly. Most tools allow some tailoring of the output. You might create a template that communicates context, followed by a summary, key decisions, and action items, or a narrative format like "Who said what, why it matters, and next steps." Experiment with formats that resonate with your clients. A concise, well-structured recap is part of your professional brand.

Keep in mind that your summary, like anything you produce that's client-facing, is part of your brand. A sharp and clear meeting recap signals to the client that you listened carefully and processed what was said. It's an opportunity to show insight and add value. In contrast, a sloppy auto-generated summary sent without review can undermine your credibility. Use AI to draft the recap, then edit it to ensure it's accurate in tone and emphasis. This extra five minutes of effort can turn a generic summary into a polished deliverable that reinforces your expertise and attentiveness.

Ethical and Practical Considerations

Responsible use of AI note-taking tools starts with transparency and respect. Before you hit "record" on an AI meeting assistant, make sure you've addressed the following:

Before Recording...

- **Ask for permission.** In many jurisdictions (and certainly many company cultures), it is customary (if not required) to get consent from meeting participants before recording. Some jurisdictions have two-party consent laws requiring everyone's agreement. Even if not legally required, it's good manners and builds trust when you ask, "Mind if I use an AI assistant to help take notes in this meeting?"

- **Disclose the tool being used.** Don't record surreptitiously. Be clear if you're using Fathom, Fireflies, etc., and explain that it will transcribe the conversation. Most people will be fine with it if you explain the purpose ("...so I can focus on our talk and not miss details"), but they should know who or what is listening in.

- **Make sharing opt-in.** Reassure participants that the transcript will be shared only with authorized people. For instance, you might say, "I'll keep the notes for my reference and can share with you afterward if you'd like." Never forward AI-generated notes to others (or especially to folks who weren't in the meeting) without consent from the meeting owner.

Storage Practices

- **Use secure tools.** Ensure the platform has enterprise-grade encryption and security. If you're discussing sensitive client information, you don't want it stored on a leaky server. Look for statements about data encryption in transit and at rest (e.g., SSL/TLS and AES-256). Many top providers offer security whitepapers; for example, some AI meeting apps tout SOC 2, GDPR, and even HIPAA compliance to assure you that data is handled responsibly.

- **Store recordings in approved locations.** Some tools save recordings and transcripts on their cloud by default. Verify where those files go and who has access. You may choose to download the transcript and save it in a company-approved secure drive, then delete it from the external tool. Always keep client confidentiality in mind.

- **Set retention policies.** Don't keep recordings or transcripts longer than necessary. Not every meeting needs to be archived forever. Deleting old records (especially audio recordings) can reduce risk. Some tools let you auto-delete data after user-defined durations. Use that feature for routine meetings that don't need long-term storage.

Limitations

No matter how advanced, AI note-takers have limitations to be aware of. For example:

- **They can miss nuance.** Sarcasm, tone, humor, or underlying emotions may not translate in a transcript. An AI-generated summary might downplay a key concern if it doesn't understand the context or if the speaker was being indirect.

- **Transcription isn't 100 percent accurate.** Heavy accents, poor audio, or industry-specific jargon can lead to mistakes. Important details could be transcribed incorrectly. Always skim critical sections of the transcript (or listen to the recording at key moments) when nuance is important.

- **Don't rely on summaries alone.** The AI's recap is meant to assist, not replace your judgment. Use it as a starting point. If something in a summary seems off, revisit the full conversation. And certainly, for high-stakes meetings, you should review the transcript or recording to ensure nothing was misconstrued.

Finally, notwithstanding the ability to auto-forward meeting summaries, be judicious when using that feature. While it's a real convenience (eliminating a step), a summary might inadvertently describe a discussion as generally positive but gloss over a client complaint that had been voiced. An inappropriately positive spin on a contentious meeting might leave your client feeling that their concern was being downplayed, and question whether you truly heard them. It's good practice, especially before you're fully comfortable with your notetaking tool of choice, to review the full transcript, catch any nuance, and follow up personally to address any client concerns. Technology can miss subtext.

Integrating Meeting Tools Into Your Workflow

The value of AI notetaking multiplies when it's embedded into your broader systems and routines. More important than just having notes, of course, is using them. Consider weaving your meeting assistant into the apps you already use and the processes you follow to derive the greatest benefit.

Example Integrations. Most AI meeting tools offer native integrations or connectors to workflow orchestration apps to streamline workflow. A few high-impact ones to consider include:

- **CRM (Salesforce, HubSpot).** Automatically tag meeting notes to client or deal records. For instance, Fireflies can log call summaries to Salesforce. This means your CRM always has the latest call info, and you don't have to duplicate data entry after meetings.

- **Knowledge bases (Notion, Coda, Confluence).** Build a searchable library of insights. You can have transcripts and summaries fed into a Notion database, for example, organized by client or project. Over time you accumulate a valuable knowledge base where you can query past conversations easily.

- **Project management (Trello, Asana, Monday).** Link action items from meetings to your task boards. If in a meeting someone says,

"I'll send the report by Friday," the AI can detect that commitment and create a task in Asana with the deadline. This keeps your to-do lists updated without manual effort.

- **Team chat (Slack, Microsoft Teams).** Post meeting outcomes to the relevant channel. Many tools can be set to drop a summary into Slack right after the call. This keeps teammates who couldn't attend in the loop and creates an easy spot for discussion or clarification on what was decided.

Beyond integrations, think about how to standardize and take full advantage of what AI notetakers provide. For example, by using templates and tags, you can develop a consistent structure for your meeting notes so that both you and your clients know what to expect:

- **For recurring meeting types (client onboarding, weekly status update, retrospective, etc.).** Use pre-defined templates if your tool offers them. For example, Otter.ai recently introduced Meeting Types with tailored summary prompts for different kinds of meetings. A sales call might highlight objections and next steps, whereas a project kickoff might highlight goals and responsibilities. By using a template, you ensure the AI is surfacing the insights you care about.

- **Apply tags or labels consistently.** If your tool or your workflow allows tagging of notes (e.g., #Decision, #Risk, #ActionItem), decide on a tagging taxonomy and stick to it. Later, you can filter or search transcripts by these tags to pull up, say, all #Risks discussed in the last month. Some systems will auto-tag certain phrases, but you can usually add your own, too.

Institutional Memory on Demand

Make a practice of storing all AI-generated call summaries in a shared workspace (e.g., Notion or a Teams channel), organized by client and project phase. Over a long engagement, you'll amass many pages of notes, providing a real-time history of the project. When your client sees how valuable this is, they're likely to start referencing it, too. If new stakeholders get involved, you'll be able to onboard them in record time by pointing them to this archive. They'll be able to follow the evolution of decisions and context behind every major change, getting up to speed in days instead of months. This further instills trust by demonstrating your firm grasp of the client's story. By integrating an AI meeting tool into a knowledge-sharing process, you'll effectively turn ephemeral conversations into lasting institutional memory.

Developing Your Listening Infrastructure

Think of your meetings not as isolated, one-off events, but as a continuous stream of insight and data that feeds your consulting practice. To maximize the benefit, you'll want a *listening infrastructure*—a system that captures, organizes, and leverages the outputs of all those conversations. Components should include:

- **Centralized storage of transcripts and summaries.** Use a single repository or hub for all notes so you (and your team) aren't hunting through emails or random folders to find them. Consistency is key.

- **Thematic analysis over time.** With all your meetings captured, what trends emerge over weeks or months? Perhaps every client in a certain industry mentions "data silos" as a challenge, providing a signal you can develop an offering around. AI can help by clustering themes or even producing periodic insights.

- **Integration with client dashboards or deliverables.** If you provide clients with a dashboard or regular report, consider feeding relevant meeting insights into it. If the client has a KPI dashboard, maybe a section for "This week's decisions" could auto-populate from your meeting notes. This keeps the important discussions connected to the metrics and outcomes.

- **Collaboration and sharing.** Make it easy to share selective pieces of transcripts or highlights with clients or colleagues. Your listening infrastructure should allow controlled access—maybe you want to share a summary but not the raw transcript (or vice versa). Some tools let you share specific clips of meeting recordings, which can be powerful. If a client was absent in a meeting where a critical decision was made, you can share just the excerpt of that moment so they can hear it verbatim.

Always consider how much of your client wisdom is being captured, and how much is evaporating after each meeting. If the answer is "not much" or "only what I scribble in my notebook," there is huge opportunity for improvement. Your insights are one of your biggest assets; don't let them vanish.

Listening system summary. As you develop your listening infrastructure, keep these known good practices in mind:

- **Use AI note-taking tools** in all key client meetings (especially external, client-facing calls where details matter).

- **Create a standard summary** format or template for your notes so they are easy to parse and have a uniform quality.

- **Store all call notes** in a searchable repository (cloud drive, Notion, CRM—wherever it will actually be used).

- **Schedule regular reviews** of transcripts or summaries. For example, spend thirty minutes each Friday to scan the week's conversations for any overlooked follow-ups or emerging issues. This habit can catch red flags before they become fires.

By checking these boxes, you're developing a powerful listening system for your practice.

What to Look for in a Tool

Not all AI meeting assistants are created equal. When evaluating options, consider the following factors in light of your specific needs:

- **Transcription accuracy.** How well does the tool handle various accents, technical terminology, or multiple people speaking? Some services boast high accuracy, but your experience may differ. If you work in an area with lots of jargon (as many of us consultants famously do), test the tool to see how it handles those terms. Also consider how it signals that it's unsure about a transcription. For example, does it mark inaudible sections or guess incorrectly without warning?

- **Real-time vs. post-call features.** Do you need live captions or live summaries during the meeting, or is it enough to get the notes afterward? Real-time features enable you to correct course *during* a meeting ("It looks like the AI missed that last point, let me restate it"). However, real-time processing can be less accurate; some teams may prefer a high-quality summary a few minutes after the meeting instead.

- **Customization of outputs.** Can you tailor the format of notes and summaries? Look for tools that let you adjust how verbose the summary is, or highlight certain sections. The ability to create custom templates or choose different summary styles is a plus. This way, the AI's output can match your consulting deliverables format. Also, check if the tool supports speaker identification naming (can you label "Speaker 1" as "Christine—Client CEO"), which makes transcripts more readable.

- **Integration with your stack.** As discussed, integration is key to embedding the tool in your workflow. If you work with Outlook and Microsoft Teams, ensure the AI assistant works well there. If your firm uses Google Workspace, check for Google Calendar, Docs, and Drive integrations. Some tools have dozens of integrations, including project management and CRM, whereas others might only offer a Zapier connector. Choose one that will save you time by automating handoffs to the other apps you already use.

- **Pricing and scalability.** Evaluate free vs. paid offerings; many tools have a free tier. Consider how the cost would scale if your usage grows or if you onboard team members. Also, check data retention in free vs. paid plans (e.g., some free plans might delete transcripts after thirty days). For enterprise deployments, ask about options for self-hosting or enterprise licenses if data control is a major concern.

Finally, remember to choose a tool that aligns with your consulting style and values. This is where the human-centered philosophy comes in. If your priority is client relationship and trust, you might value transparency features and simplicity over a tool that's hyper-automated but opaque. If you are a solo consultant who values cost-efficiency, a free or low-cost tool with slightly fewer bells and whistles might serve you better than an expensive enterprise platform. If you're tech-savvy and love to tinker, you might enjoy a tool with a lot of customization and integration options; if not, pick one with a straightforward UI that works for you. The best tool is the one that you will actually use consistently.

Listen Better, Deliver Smarter

AI meeting tools won't replace your insight or expertise, but they can certainly enhance it. Used intentionally, these assistants help you stay more present in conversations, capture nuance for later review, reduce the prep time between sessions, and create client-ready outputs faster. In essence, they allow you to listen better *and* longer. You're no longer constrained by the limits of human notetaking or memory. The AI extends your listening capacity to cover every word, while you devote your mind to higher-level understanding.

By embracing these tools in a thoughtful way, you turn moments into memory, memory into insight, and insight into advantage. You become the consultant who never misses a detail, always follows up, and continually learns from every client interaction. That ultimately means delivering smarter recommendations and coaching your clients more effectively. The key takeaway is that augmented listening is a superpower any consultant can develop—one that strengthens your human connection with clients while harnessing the best of AI. It's about *hearing* more, so you can *give* more.

ROB BERG

CHAPTER EIGHT

High-Impact AI Tools for Consultants

Not every valuable AI tool fits neatly into a single box, but that doesn't mean it should be overlooked. In this chapter, we highlight a range of versatile, high-impact tool categories that consultants can use to conduct research, analyze data, enhance visual storytelling, and repurpose content. Think of this as a curated sampler of AI-enabled leverage points across your consulting practice. The focus here is not on endorsing specific tools (platforms evolve quickly), but on understanding the types of leverage each category can provide so you can evaluate, experiment, and evolve your own toolkit accordingly. Embracing these tools isn't about replacing the human element of consulting; it's about freeing you to amplify your uniquely human strengths (insight, creativity, empathy) in service of your clients.

As such, throughout this chapter, we'll stay grounded in a human-first consulting mindset. That means using AI as an augmentation to your expertise, not a substitute for it. Each tool category offers ways to work smarter or faster, but it's up to you to apply judgment, authenticity, and creativity. The most successful AI-enhanced consultants are intentional, meaning they select tools aligned with their personal consulting style and their clients' needs, rather than chasing every new trend. With that idea in mind, let's explore four major categories of AI tools and how they can transform your workflow.

AI-Driven Research Assistants

Research is foundational to almost every consulting engagement, whether it's scanning the competitive landscape, summarizing academic evidence, or prepping for a client pitch. AI-powered research assistants are designed to compress the time between question and answer, while increasing your confidence in the results. In essence, they leverage AI's speed and breadth of knowledge to augment your fact-finding capabilities.

Key Benefits. These tools fundamentally change how you gather and synthesize information:

- **Accelerated synthesis of large knowledge bases.** An AI research assistant can sift through millions of webpages or papers in minutes, giving you distilled answers that would take a human many hours to assemble.

- **Embedded citations and traceability.** Unlike a generic web search, many AI research tools provide sources for every claim (though it is *hypercritical* to check every source for accuracy), so you can verify facts and trace them back to origin. This builds credibility in your deliverables.

- **Reduced manual search and scanning.** Instead of poring over dozens of search results and PDFs, you can ask a direct question and get a concise, referenced summary. This reduces information overload and lets you focus on analysis and application.

In practice, suppose you're preparing a briefing on regulatory trends in the fintech space. Traditionally, you might spend half a day on Google, opening fifty tabs and reading multiple reports to piece together the latest developments. With an AI research assistant, you could pose a question, such as, *"What new fintech regulations emerged in Europe in the past year?"* and receive a well-structured summary of the key updates, each with a citation to the source. In fifteen minutes, you have a first-draft outline complete with references to official regulatory texts and news articles. That's not the end of your research (you'll still review and interpret the findings), but it's a running start that saves hours and ensures no major point is missed.

Representative Tools. A number of AI research assistants have emerged to tackle different knowledge domains and use cases. For example:

- **Perplexity AI** is an AI-powered search engine that answers user questions (even about current news events) with conversational summaries and a list of relevant sources. It's especially handy for business intelligence on current topics, because it retrieves real-time information and cites multiple web sources for verification.

- **Consensus** is a research tool focused on scholarly and scientific literature. It can sift through more than 200 million academic papers to find evidence-based answers, providing insight into the consensus emerging from peer-reviewed studies. This is ideal for consulting projects in healthcare, science, or public policy, where clients expect recommendations grounded in rigorous research.

- **Scite.ai** is a platform that adds nuance to academic findings by showing how each paper has been cited by others. It uses AI to indicate whether subsequent studies support or contradict the claims of a given paper. For a consultant, this helps quickly identify which research findings are robust vs. controversial, ensuring your advice rests on well-vetted evidence.

These tools are especially valuable when you need to quickly assess what research literature said about a particular topic relevant to your project work—say, for example the effects of electronic employee monitoring on productivity. Instead of manually searching PubMed or EBSCOhost, you can use Consensus to query *"Does the use of employee monitoring software by organizations improve employee productivity?"* The AI tool will then return a summary of findings from relevant peer-reviewed studies, complete with citations and even a brief indication of overall agreement levels among the papers. In this instance, you might discover that while some studies support improved productivity, the majority (consensus) indicate detrimental impacts to productivity and employee morale. To validate further, you might use Scite.ai to examine an especially pivotal study that later papers overwhelmingly cited with evidence supporting the consensus view. Armed with this AI-assisted research, you can confidently brief your client with evidence-backed insights. Something that previously might have taken days of literature review could be accomplished in a few hours. The key point is that the AI assistants don't replace your expertise (you still need to interpret which studies are most relevant and applicable), but it can dramatically accelerate your ability to gather credible information.

AI for Data Analysis and Visualization

Consultants rely heavily on data, but such data is rarely presented in nicely formatted, easy-to-interpret dashboards. Often, the real work is making sense of messy spreadsheets, uncovering patterns in the numbers, and packaging insights in client-ready visuals. AI tools in this category shift the balance by automating many aspects of data analysis and visualization. They empower consultants who aren't data scientists to extract value from data quickly and confidently.

With the right AI tools, you no longer need to write complex code or formulas to interrogate data. Whether you're exploring KPIs in a client's Excel workbook or building a high-level financial model, new AI-driven data tools let you ask plain-language questions and get structured answers. Imagine opening a massive CSV export of sales data and simply typing, *"Show me in a bar chart which product categories grew the fastest last quarter."* In seconds, an AI assistant can read the data, perform the calculations, and produce a ready-to-use chart highlighting the growth rates by category. This natural language interface lowers the barrier to analysis, allowing you to iterate on ideas without lengthy manual work. The leverage here is turning raw data into insights (and visuals) with minimal friction.

Key Use Cases. Data-focused AI tools can assist at multiple stages of your analytic workflow:

- **Cleaning and structuring data.** They can automatically detect inconsistencies, fill missing values, or normalize formats in seconds—tasks that might take hours of Excel wrangling.

- **Exploratory analysis.** Ask questions of your dataset and get answers in the form of summaries, tables, or charts. For example, *"What were the outlier transactions last month?"* could yield a quick list of anomalies.

- **Generating charts and dashboards.** With a simple prompt, you can create graphs or even entire dashboards. The AI handles the coding or chart setup, so you can focus on interpreting the visual.

- **Summarizing trends and anomalies.** Instead of manually writing commentary, you can have the AI scan the data and draft bullet points like, *"Sales spiked in the Northeast region in July, which were 22.7% higher than June, then returned to average levels with a 20.2% decrease in August"* with the evidence to back it up.

Representative Tools. A few examples illustrate the range of capabilities here:

- **ChatGPT (Advanced Data Analysis module).** OpenAI's ChatGPT offers a mode (formerly called Code Interpreter) that lets you upload data files (spreadsheets, CSVs, JSON, etc.) and then interact via natural language to analyze them. Under the hood, it writes and executes Python code. This means you can prompt it to calculate statistics or generate visualizations, and it will produce the results directly in the chat. It's designed to perform complex data tasks (from crunching numbers to plotting charts) through simple prompts, making data analysis accessible to those without coding expertise. For instance, it can securely execute Python to do things like statistical analysis, data cleaning, or chart creation, and then show you the output right inline. As a result, analyzing a large dataset can feel like having a conversation with a data-savvy analyst who can also draw.

- **Spreadsheet assistants (Microsoft's Copilot in Excel and Google's Gemini for Workspace).** The major productivity suites have embedded AI copilots directly into tools like Excel and Google Sheets. Microsoft's Excel Copilot can generate formula columns, highlight trends, and even suggest charts or PivotTable insights just by asking or clicking a cell. For example, you can prompt, *"Analyze this data for any seasonal patterns"* and it might create a summary or a visualization of trends over time. Google has also embedded its Gemini AI in Google Sheets, which offers similar capabilities. It can create tables, write formulas, generate analyses and charts, and even summarize datasets in natural language. The advantage of these integrated assistants is that they work within tools you already use, enhancing your existing workflow.

- **AI-Powered BI platforms (ThoughtSpot, Tableau GPT).** A new wave of business intelligence tools allows you to upload a dataset and then automatically generate dashboards or reports. For example, ThoughtSpot transforms spreadsheet data into interactive dashboards through natural language search, automatically visualizing key metrics. Similarly, established BI software like Tableau and Power BI are adding GPT-powered chat interfaces where you can ask, *"Show a pie chart of revenue by region"* and it builds it for you. These tools remove a lot of the tedium in data visualization, letting you iterate on insights faster.

These tools are especially useful for rapid data analysis. Imagine being tasked with analyzing several years of sales and pricing data for a client's product lines. Using traditional methods, you might spend days writing Excel formulas or SQL queries to find patterns in price elasticity. By uploading the data into ChatGPT's Advanced Data Analysis feature and asking a series of questions in plain English, such as, *"Which customer segments showed an increase in revenue when prices were raised, and which segments saw a decrease?,"* the AI will quickly parse the dataset and identify those segments where price increases impacted unit sales—and even generate a simple chart highlighting which segments were adversely impacted and which were price-insensitive. You might then ask the AI to describe any identified trends in a few bullet points to produce a draft summary you could refine. In the end, you'd have an analysis and a visualization ready in a fraction of the time it would normally take, allowing you to focus on crafting strategic recommendations around the findings.

AI-Based Presentation and Design Tools

How you package insight often determines whether it sticks. In consulting, delivering value isn't just about the analysis; it's also about the story you tell and the professionalism of your deliverables. AI design tools help bridge the gap between great ideas and great presentations/documents by automating

146

parts of the design and writing process. They give consultants a "designer on demand," ensuring the work looks polished and on-brand without consuming countless hours in PowerPoint or Adobe Creative Suite.

These AI tools offer leverage in a few key ways. They can generate presentation drafts from a simple prompt or outline, suggest visuals and layouts that match a desired style, and even adapt designs to fit a company's branding guidelines. The goal is to let you iterate on the content and narrative, while the AI handles time-consuming formatting and design choices. AI design tools help consultants:

- **Generate drafts from an outline.** Provide a rough outline or a few key points, and the AI will produce a set of slides or a document with fleshed-out sections, placeholder visuals, and suggested talking points. It's like magic when facing a blank page.

- **Match branding and tone.** You can usually specify the style (e.g., *"Make it look like a Big Four style report"* or upload brand colors/logo) and the tool will apply consistent fonts, colors, and imagery. This means even non-designers can achieve a professional, on-brand look.

- **Rapidly iterate storyboards.** Because it's so quick to generate versions, you can experiment with different story flows or design metaphors. The AI might propose a timeline graphic or a framework diagram that sparks your creativity, which you can then refine.

Developing a Presentation

Let's say you're building a strategy deck and facing a tight deadline. You have your analysis results and some rough insights, but not much time to craft the narrative flow or design slides from scratch. An AI presentation tool can be a lifesaver. You could input something as simple as a one-line prompt like, *"Retail expansion strategy for Company X—slides covering market analysis, SWOT, entry plan, financial projections."* Within minutes, the AI generates a first draft of a ten-slide deck. It might contain a title slide with the company's logo and a suggested tagline, an agenda, a SWOT matrix (populated with plausible points to refine), a market trends slide with icons, and so on. All slides come out in a clean layout with a coherent visual theme. Now, this draft isn't client-ready on its own, but it gives you a scaffold to build on. Instead of starting from zero, you spend your time reviewing and editing the AI's suggestions, injecting your specific insights, ensuring accuracy, and adjusting the narrative. In this example, the AI shortened the distance between idea and draft dramatically.

Representative Tools. Several innovative platforms exemplify what AI can do in the realm of presentations and design:

- **Gamma.** An AI tool specifically geared toward creating slide decks and visual documents from text-based input. You can literally paste an outline or bullet points, and Gamma will convert it into a professional-looking deck with slides for each point, relevant imagery or icons, and a cohesive design theme. Users have noted that Gamma's generated slides are often surprisingly close to polished; you might actually consider presenting them with only minor tweaks. It exports to PowerPoint or PDF as needed. This is like having a first-draft deck creator on call; you focus on the outline of your story, and Gamma handles the initial composition of slides.

- **Canva.** Canva is a popular design platform, and its Magic Design feature injects AI into the mix for creating presentations and graphics. You can describe the kind of visual you need (or provide content like images or text), and Canva will suggest customized layouts and template designs based on your input. For example, feed it the text for a slide or upload a couple of images, and it will generate a variety of slide design options with different arrangements, styles, and fonts for you to pick from. It's like an AI art director that provides you with multiple comps in seconds. Canva's AI design assistance will also make automatic adjustments, ensuring alignment, proper spacing, and color harmony, so that the end result looks like it was made by a pro designer.

- **Tome.** A "narrative-first" presentation builder, Tome uses AI to help structure a compelling story flow and generate written and visual content to support it. It's great for storytelling decks (like pitch narratives or concept proposals). You might input a short description of

148

the story you want to tell, and Tome will produce a sequence of slides where each slide has a narrative point and an image or graphic to illustrate it. Its AI might draft some copy and then suggest an image (using generative image AI for unique visuals). Essentially, Tome blends copywriting and design; you get both text suggestions and visual suggestions in one tool. This can really help if you struggle with what to say on each slide or how to visualize an idea.

Using Gamma

When preparing a proposal deck for a prospective client, you want it not only to capture the ideas in your head, but to look good as well. Using an AI tool like Gamma, you can simply type a few key prompts about your client's situation and your proposed solution and produce an instant deck that includes a cover slide with the client's logo, an outline of sections, and even a few illustrative icons on slides discussing critical areas of interest (e.g., customer experience, operating efficiency, etc.). While the auto-generated text on the slides can seem somewhat generic, it provides a starting point and a structure for what and how to create the presentation. It's up to you to spend additional time customizing the slides, perhaps by replacing generic points with client-specific insights and tweaking the visuals. In short order, you can create a proposal that's visually compelling and tailored to the client with the initial design, structure, and content produced in minutes rather than hours fumbling with the right approach. I find this to be an exceptional way to remove the writer's block that often precedes those tedious proposal-writing sessions when you're staring at a blank page waiting for inspiration. AI provides that inspiration (and considerably more) to get you moving quickly in the right direction.

That said, AI won't (and shouldn't) replace your consulting judgment or storytelling ability; you still decide *what* goes on each slide, but it can drastically reduce the time spent on messing with fonts, layouts, and stock images. In high-stakes situations where timelines are tight and first impressions count, having AI tools like Gamma accelerates a formerly slow, iterative design process. The result is more time to focus on the message and less anxiety about making it "look pretty."

Audio, Video, and Content Repurposing Tools

If you create content—whether for clients (e.g., training materials, explainer videos) or for your own marketing (blog posts, podcasts, webinars)—AI can help you make more of what you already have. Content repurposing tools use AI to transcribe, edit, and transform content from one format into others with minimal effort. This category is all about leverage through multiplication—turning one piece of content into many derivatives, and polishing media content without specialized skills.

Consider a common scenario where you host a thirty-minute webinar sharing insights from a project. Traditionally, turning that into other assets would be a heavy lift (writing up a summary, editing clips for social media, maybe recording a follow-up voiceover). AI tools can now do much of that grunt work. They can transcribe the audio, identify key takeaways, cut filler words, generate summaries, and even create new content like social posts or blog drafts based on the transcript. Similarly, if you film a talking-head video or record a podcast, AI tools can clean up the audio and help you redistribute that content in multiple channels quickly. In addition, these tools can help with a variety of other formerly time-consuming tasks. For example:

- **Polishing raw recordings.** You record a Zoom meeting or a training session. AI tools can remove all the "um's" and long pauses, fix minor audio issues, and even trim the video to remove dead air, giving you a cleaner version to share.

- **Transcribing and summarizing.** Any audio or video can be transcribed to text automatically. Beyond that, AI can summarize the transcript or extract key points (great for meeting notes or creating an executive summary of a presentation).

- **Transforming content.** Turn a webinar into a blog post, a podcast into a series of tweets, or a slide deck into an annotated video. AI can generate these new formats by understanding the source content and rewriting or reformatting it appropriately.

- **Creating synthetic media.** Need a voiceover recorded or a paragraph of text turned into spoken audio? Modern AI voice generators can produce lifelike voice tracks, so you can add narration to a video or generate an audio version of a report without hiring voice talent.

Representative Tools. To illustrate the power of this category, let's look at a few tools and what they enable:

- **Descript.** A game-changing tool for editing audio and video as easily as a text document. Descript automatically transcribes your audio/video, so you edit the transcript to edit the media. Delete a sentence in the text, and that portion of the audio/video is cut out. This makes removing filler words ("um," "uh," etc.) or mistakes as simple as hitting delete on the text. It also has an overdub feature; if you need to add a word, you can type it and it will synthesize it in your voice. For consultants, Descript is fantastic for cleaning up recorded presentations or creating snippets. For example, you could record yourself explaining a concept, then easily cut out tangents or long silences, yielding a tight two-minute clip for LinkedIn. No complex timeline editing required—just cut the text. Descript essentially gives you a quick audio/video editor that anyone can use, with AI helping to seamlessly patch the edits.

- **Castmagic.** A specialized AI tool for podcasters and content creators that auto-generates a trove of content from an audio recording. Upload an audio file (podcast episode, webinar audio, etc.), and Castmagic will produce the full transcript, a succinct summary, show notes, a list of key highlights or timestamps, and even suggested social media posts quoting interesting bits. It's like having a team of assistants who listen to your recording and prepare all the ancillary content. For a consultant, this can be gold. Imagine you record a twenty-minute panel discussion on industry trends. Castmagic can give you a ready summary to send to your client, plus five short social posts pulling quotable insights (which you can refine and post to demonstrate your thought leadership). It saves you from re-listening and manually writing summaries.

- **ElevenLabs.** An AI voice generator that can produce lifelike audio from text. It uses advanced text-to-speech models to read any script in a natural, human-sounding way. With multiple voices and tones to choose from, you can create professional voiceovers without recording anything yourself. For example, if you have a written case study and want to offer it as an audio reading on your website, you can use these tools to generate it. The quality is far beyond the robotic voices of the past; listeners may not even realize it's AI. Consultants can leverage this by adding audio narration to slide decks (make a self-narrated video version of a proposal), creating spoken versions of reports for busy execs, or even personalizing outreach with a quick voice message that you typed out and had an AI speak.

Here's a typical use case that leverages these tools: Suppose you've just finished a lengthy Zoom workshop with a client's team walking through a new organizational framework, and wish to make the content available in different formats for the workshop participants' reinforcement. You can take the Zoom recording and feed it into Descript and, in minutes, produce a transcript. You can then use Descript's AI to automatically remove filler words and awkward pauses, as well as remove off-topic sections by simply cutting the relevant text. You might then export an audio-only version of the cleaned workshop and upload it to Castmagic, where you can generate a summary of the workshop, produce a set of concise bullet-point notes, and even a draft follow-up email to send to the participants with key takeaways derived from the session audio. It could also pull a couple of memorable quotes from the discussion to use in a newsletter, email, or other communication. In about thirty minutes of work using these AI tools, you can transform a one-time meeting into a polished video, a written summary, a set of quotes, and an email update. The content can live far beyond the initial workshop, multiplying its impact without having to manually transcribe or tediously edit media files.

Think about any content you've created in the past ninety days. It could be a client presentation, a webinar, a research report, or even a series of insightful emails. Could one of these AI tools help repackage, re-share, or reuse that content in a new format? Consultants often sit on valuable intellectual property that only reaches one audience in one format one time. With AI, you might take a slide deck and turn it into a blog post, or take a research report and create a podcast-style audio reading of it. What value could you unlock by giving your existing content new life?

How to Choose High-Impact Tools

Every week brings new AI announcements, and it's easy to feel over-whelmed or distracted by shiny new features. As we've seen, the possibilities are exciting—but remember that effective consulting isn't about having *every* tool, it's about using the *right* tools to deliver better outcomes. How do you navigate the noise and choose high-impact tools for your own prac-tice? A smarter approach is to be purposeful and aligned with your unique context.

- **Focus on use case first.** Identify the specific task or workflow you want to improve. Are you trying to speed up research? Analyze data faster? Create better visuals? Start with a clear problem or oppor-tunity in your work where AI could help. This use-case mindset pre-vents adopting technology for its own sake. For example, if you never produce audio/video content, you probably don't need a tool like Castmagic, but you might desperately need a research assistant. By focusing on your pain points or high-value activities, you'll choose tools that actually make a difference.

- **Align with your workflow and style.** Favor tools that integrate into how you already work or that feel natural to use. If you live in Mi-crosoft Excel, using Excel's built-in Copilot might beat adopting a separate data analysis platform. If you prefer writing outlines in Word, a plugin or tool that turns outlines into slides will fit you bet-ter than one requiring a completely new interface. Also, ensure the tool aligns with your consulting identity. If your brand is all about customized, high-touch deliverables, you'll want AI that enhances that (perhaps by saving you time to add personal touches) rather than a tool that makes everything feel templated. In short, choose AI that amplifies your authenticity, rather than compromising it.

- **Favor stability and simplicity over novelty.** In the fast-moving AI market, not every new tool is production-ready. When bringing something into client-facing work, reliability matters. It can be better to use a well-supported feature in a major platform (even if it's not the absolute cutting edge) than a beta-stage app that might glitch or disappear. Also, simpler tools that do one thing well often perform better than complex, multi-purpose platforms. Especially in front of clients, you want tools that won't fail you and that you're comfortable using. Build confidence with a tool in low-risk scenarios first, then elevate it to high-stakes use once you trust it.

- **One step at a time.** It's tempting to sign up for a dozen AI tools at once. Resist that urge. A better strategy is to pick one promising tool in a category and genuinely learn how to use it in your routine. For instance, spend time really exploring an AI research assistant on your next project and see what it can do. Building that muscle memory and integrating it into your process is crucial to seeing the benefit. Once it's second nature, then consider adding another tool to the mix. You'll get more value mastering a few versatile tools than dabbling superficially in many. Remember, high impact comes from depth of use, not breadth of tools.

To bring this together, here's a simple way to think about building your AI-augmented consulting stack: identify the key categories of work you do, note what you currently rely on, and then pick a high-impact AI tool in that category to pilot. The table below illustrates this approach:

Category	Current Tool	One to Explore
Research	Google Search	Perplexity (AI Q&A Search)
Data Analysis	Excel spreadsheets	ChatGPT Advanced Data Analysis
Visual Storytelling	PowerPoint slides	Gamma or Tome (AI deck drafting)
Content Repurposing	*None / manual work*	Descript + Castmagic (AI editing & summarizing)

You don't need to master twenty different tools to be successful. Instead, focus on the two or three categories that map most closely to the activities

154

you perform frequently that clients truly pay you for—and get really good at using AI to elevate those. If research and slide-writing are where you spend most of your time, turbocharge those with an AI assistant and an AI slide generator. You'll likely feel the impact in time saved and enhanced quality immediately. It's far better to deeply integrate a handful of high-impact tools (so they become second nature in your workflow) than to superficially try everything new under the sun.

Think Beyond the Obvious

The AI tool categories we explored in this chapter aren't always part of the traditional consulting toolkit, but they're often the difference between working hard and working smart in the modern era. By now, you should see that each category offers a form of leverage. Faster research, deeper data insights, quicker content creation, and broader content reach are all enabled using these tools. As you consider adopting them, start with a specific deliverable or process you handle frequently. Perhaps it's a monthly industry research brief, a recurring KPI dashboard, a standard proposal deck, or routine meeting follow-up reports. Ask yourself, "Could one of these AI tools cut my production time in half? Could it make the output twice as good for my client?" If the answer is yes, that's a strong signal to experiment.

Ultimately, the consultants who thrive in our AI-enhanced world won't be those who simply use the most tools' they'll be those who curate the right tools that align with their own style, workflow, and client promises. High-impact tool use is about *intention*. It's the consultant who says, "I stand for quality insights and personalized service, so I use AI tool X to give me more time with the client and AI tool Y to ensure no detail is missed in my research" who will shine. This aligns with the core theme of authenticity; use AI to amplify your strengths, not to chase someone else's formula.

As you expand your toolkit, maintain humility and curiosity. AI will keep evolving, and there will always be something new around the corner. Approach each new capability with the mindset of a learner. Consider how a given tool might help you to deliver more value, and always maintain a degree of healthy skepticism, considering whether a given tool truly enhances your work or just adds complexity. By doing so, you'll build an AI-enabled practice that is resilient, distinctive, and very much human-centered. Embrace the leverage these high-impact tools offer, but never outsource the core of the judgment, creativity, and trust-building with clients that make your consulting valuable. Keep those at the forefront, and let AI do the heavy lifting in the background. In the next part of this book, we'll dive into detailed guides and use-cases for many of the tools mentioned, so you can translate this understanding into hands-on action. For now, start thinking boldly about how you can reinvent parts of your work with a little AI assistance, and get ready to chart your own path in this new landscape of consulting. The tools are here; the next move is yours.

Part 3

Playbooks & Prompts

With mindset set and tools in hand, Part 3 is where it all comes together. This section offers practical blueprints for applying AI across the full arc of your consulting work—from prospecting and proposals to project execution and thought leadership. Each chapter is a deep dive into real consulting use cases, equipped with adaptable frameworks, working templates, and curated prompt libraries you can tailor to your own style and client base.

Think of this as your consulting lab. We'll explore how to co-create service offerings with AI, accelerate research and writing, streamline client interactions, and master prompting as a strategic skill instead of just a technical one. These playbooks aren't rigid recipes. They're designed to spark ideas, invite customization, and expand your sense of what's possible. Whether you're building your first AI-powered service or refining an established workflow, this is where strategy turns into practice—and practice turns into performance.

ROB BERG

◆

CHAPTER NINE

Sales and Marketing Use Cases

We've had a glimpse into how AI can transform the way consultants *deliver* value, so we'll next dive into how it can transform how they *market and sell* that value. In this chapter, we explore how consultants can apply AI to streamline outreach, create personalized content, identify market trends, and grow visibility with less effort and more precision. These aren't abstract capabilities; they're concrete, practical ways to turn AI from a curiosity into a lead-generation and relationship-building engine. Importantly, this chapter doesn't suggest outsourcing your marketing voice to a machine. Instead, it offers strategies for consultants to stay visible, relevant, and responsive with AI as a copilot that reinforces authenticity and human connection.

Prospecting with AI

Finding the right prospects at the right time is the foundation of business development. AI agents and assistants can amplify your reach and relevance by automating the grunt work of research and data gathering. For example, they can continuously scrape and summarize public data about companies and decision-makers, flagging intent signals like new funding rounds, executive hires, or press mentions. This means you can spot opportunities the moment they arise. For example, an AI data agent might identify how specific types of AI-powered targeting results in dramatically higher engagement and high-probability prospects.

Concretely, imagine an AI agent that monitors Crunchbase, LinkedIn, and industry news for companies matching your ideal client profile. When it finds a match, it scrapes key facts (e.g., company size, tech stack, recent news) and compiles a snapshot. It might then draft a personalized introduction or even generate a short list of the best contacts. You, as the consultant, simply review the suggestions and decide how to proceed. By continuously running this background process, the AI never sleeps; you always have fresh, relevant leads without manually scouring hundreds of websites.

Modern tools make this practical. For example, Clay.com allows you to build automated workflows that pull in firmographic and technographic data, enriching leads dynamically. Similarly, platforms like Apollo.io offer huge B2B contact databases combined with sequencing tools, and can be paired with AI (e.g., ChatGPT) to draft smart emails. For social media, scraping tools like PhantomBuster or "browse-capable" GPTs can extract LinkedIn profile data at scale. In practice, combining these tools can yield quick wins. For example, you might use a data-enrichment tool plus ChatGPT to target fifty companies posting new data analyst jobs, then generate three email variants per persona. The result? Five qualified intro calls in a single week. This shows how AI can turn a time-consuming targeting task into fast, actionable lead lists.

These AI-driven workflows make lead generation proactive and continuous. Instead of hoping for inbound inquiries, you flood the top of your funnel with prospects precisely matched to your services. The result is more leads and *better* leads—those with a real, recent need. By combining AI data agents with a clear ideal client profile, you can stay one step ahead of competitors.

Personalized Outreach at Scale

In today's B2B world, impersonal blasts fall flat. Most recipients simply ignore cold calls, and the vast majority of people never respond to email (do you?), especially when it feels generic. What works instead is *relevance*. AI enables a new middle ground between one-off emails and generic spam in the form of semi-automated, highly personalized outreach.

AI can draft first-pass messages tailored to context. For example, say one of your leads just secured a $5 million funding round (a common trigger event). An AI prompt like, *"Write an email to the VP of Operations at [Company] mentioning their recent funding and connecting it to our analytics services. Use a professional, conversational tone"* can produce a draft email that weaves the trigger event into a problem statement. You then refine it for authenticity, perhaps adding a personal anecdote or adjusting the tone before you send. AI can also spin up multiple A/B variants automatically. In practice, testing shows that personalization works; personalized subject lines are known to boost open rates and emails tailored to the recipient's industry or role drive far higher engagement. It's not a stretch to say that personalized calls to action can double the conversion rates of generic ones.

A human-in-the-loop workflow is key. A typical process might involve the AI (e.g., ChatGPT or Claude) drafting an email based on a lead's profile and recent signals. You, the consultant, review and adjust for tone, ensuring your voice and values shine through. Then you queue the final message in an outreach platform like Apollo. After sending, the AI "listens" for the response. If the prospect opens but doesn't reply, it can suggest a follow-up; if they reply positively, it can draft a personalized thank-you message or next-step email. This keeps your campaign adaptive and responsive without manual babysitting.

Running multi-channel campaigns (e.g., email, phone, LinkedIn) can meaningfully increase meeting booking rates. AI can orchestrate these channels, too. For example, if a lead interacted on LinkedIn but hasn't replied to email, the AI might suggest a friendly LinkedIn message referencing the email. Tools like PhantomBuster can automate LinkedIn outreach via connection requests, profile comments, etc., which you can combine with AI-crafted messages. The key is persistence with relevance; AI ensures no good lead falls through the cracks, and every touchpoint feels thought-through.

Prompt engineering matters. For outreach, you might maintain a prompt library matching different personas and tones. For example:

> *Act as a consultant reaching out to [Job Title] at [Company]. They just [trigger event]. Write an email that*

connects their challenge to our solution, in a [friendly/ professional/conversational] tone.

Feeding this prompt to a large language model (LLM) generates a draft you can tweak. Over time, your prompt library grows to cover a wide variety of scenarios, such as funding announcements, new hires, press mentions, or industry trends. This systematic approach, mixing automation with your review, allows you to reach out to large numbers of leads (a multiple of what you can do without automation) without sounding robotic.

AI Agents for Engagement Follow-Up

The real test of prospecting success is what happens after a lead expresses interest. Some studies show that responding within a minute can boost conversion rates significantly (some claim higher than 300 percent!), whereas waiting even ten minutes dramatically drops the chance of engagement. Yet many businesses fail here, often not contacting even the hottest leads at all. Indeed, I'm constantly amazed when I complete a web form requesting that someone contact me—sometimes for services worth thousands of dollars— and receive no response. This has happened many times, and no organization is immune. Large and small companies are guilty here. AI can fill this gap with agents that can read, interpret, and respond thoughtfully to prospective client inquiries, ensuring your prospects get immediate attention.

Practically speaking, if a prospect downloads your latest white paper or completes a contact form at midnight, an AI agent (built with tools like n8n or Make) can immediately parse the submission, enrich the new contact (perhaps finding their LinkedIn or company info via an API), create a record in your CRM, and draft a personalized follow-up email. Because the download triggered it, the email might say, "Hi [Name], I saw you downloaded our white paper on [Topic]. I thought you might also find [Related Resource] useful. I'm curious about what insights you found most interesting." Even if the lead is half-asleep, your email arrives as their interest spark is hot. You get the ball rolling while they're still thinking about the topic.

The AI doesn't stop there. If the prospect doesn't reply in a few days, it automatically schedules a gentle nudge. If they click any links or show signs of continued interest, it flags them for an urgent follow-up. If they do reply, the AI can suggest next steps, like booking a call via a scheduling tool, inviting them to a webinar, or even looping in a colleague with relevant expertise. All this happens with minimal intervention; you oversee the workflow, but the AI does the repetitive tracking and timely nudges.

The benefit is consistency and speed. In B2B sales, the odds drop quickly if you delay. By automating your alert and follow-up system, you ensure *every* engaged lead hears from you promptly. This "white glove" follow-up—offered instantly and personally—dramatically improves conversion rates while freeing you to focus on the actual sales conversation when it happens.

Content Creation and Repurposing

Consultants sell expertise and trust, and publishing original content is one of the most effective ways to build both. Yet creating meaningful content that represents thought leadership (e.g., in blogs, LinkedIn posts, newsletters, or white papers) can be extremely time-consuming. AI lets you capture, amplify, and publish critical insights far more efficiently, acting as a force multiplier for your ideas.

Start with your own expertise. For instance, after a client meeting or a compelling idea strikes, you might record a two to three minute video or audio memo sharing that insight. AI transcription tools like Descript or Otter.ai can convert your talk into text, giving you a clean draft of your message. From there, a generative AI model such as ChatGPT, Gemini, or Claude can turn that single insight into a variety of formats, including a LinkedIn post, an email newsletter blurb, several tweet threads, and even a blog outline or draft. This expanded output reaches different audiences without much extra effort on your part.

This process offers a major win by maximizing your reach. A LinkedIn thought piece can be republished in an industry newsletter; a blog post can be sliced into social snippets; audio or video content can be transcribed or chunked into compelling sound bites. Imagine a single afternoon brainstorming session yielding content for a month's worth of marketing.

Trust and authenticity remain central. Research has shown that B2B buyers find thought-leadership content more trustworthy than standard marketing materials. In other words, the raw insight of a consultant, translated into content, carries real influence. AI simply helps package that insight in more places. It's crucial that you review and personalize all AI outputs by adding your signature style, double-checking facts, and ensuring the tone feels like *you*. But by handling the bulk of drafting, AI frees you to focus on the creative part of content—the human part—and on engaging with readers' responses.

After drafting, use scheduling and distribution tools to post across channels (for example, Buffer, Hypefury, or Typefully for social media; your email platform or Content Management System API for newsletters). Remember to maintain a library of prompts that reflect your brand voice (so ChatGPT knows to use your style) and always do a final human edit. If you do publish AI-assisted pieces, you might include a brief author's note or disclaimer (as I've done for this book), but this is optional. The goal is consistency and authenticity, not deception.

Automated Blogging with Guardrails

Beyond social posts, AI can help you maintain a steady blog without extra hiring. For example, you capture a blog idea in a simple Notion form or similar intake system, noting topic, desired length, and tone. A scheduled AI job then takes that input and generates a 600-to-800-word draft. You then review it quickly, modifying the tone or adding anecdotes to ensure it's on-brand. Once you're satisfied with the human edit, it can auto-publish to your blog site and even send it to your email or subscriber list.

The key here is guardrails. Keep a prompt library that encodes your usual structure and style (e.g., *"Write an introductory paragraph that starts with a question, include two bullet lists, end with a call to action"*). Always fact-check and localize any data or examples the AI adds (remember, models can hallucinate). Consider adding a brief note at the top of each AI-assisted post ("This article was drafted with the help of AI and edited by [Your Name].") to be transparent.

With these safeguards, consultants can produce content with less effort at greater frequency, for example publishing weekly instead of quarterly. Such consistency builds an audience and improves traction with various search and distribution algorithms. In practical terms, AI-backed blogging isn't about lazy writing; it's about accelerating the distribution of your genuine insights into the market. You still supply the knowledge and oversee the outcome. AI simply handles the routine legwork so you can stay in the flow of thought leadership.

Trend Monitoring and Market Intelligence

Staying visible and relevant also means staying informed. Clients value consultants who are up on the latest industry news and competitor moves. Again, AI can help you and your clients to be the first to know without drowning in information.

Set up AI-powered monitoring workflows that digest the news. For example, you might use Google Alerts to filter news to an AI agent for summarization and distribution. Each morning or week, the AI agent could compile bullet-point digests of recent news in your niche or client industries. It might watch competitor press releases, hiring announcements (e.g., on LinkedIn Jobs), or technology trends. Some tools can proactively alert you to emerging topics related to your keywords. The AI then distills this into an email or Slack message with highlights.

Imagine every Monday starting with a short AI-generated brief that includes three bullet points on last week's biggest industry moves, a note on a competitor you hadn't been watching who just made a pivot, and perhaps an idea suggestion. With minimal effort, you stay on top of your world and help your clients to do the same. This continuous intelligence loop not only informs client conversations (you might share the brief or schedule a quick call to discuss a major development), it also feeds your content pipeline. Many AI monitoring setups can even suggest a content idea or social post based on trending news.

The payoff is twofold: you'll appear unusually well-informed to clients, and you can tailor your service proactively. For instance, if your intelligence monitoring flags that a client's biggest competitor just raised a funding round, you can immediately advise your client on its implications (and maybe send an appropriate congratulatory note via the outreach workflow). Or if a new regulation is pending, you can prepare a blog post or checklist to address it. Being an early carrier of valuable information builds trust and positions you as a strategic partner instead of just another vendor.

The AI-Enhanced Sales Funnel

Consider traditional sales funnel stages that might include awareness, interest, consideration, decision/closed won or lost, and imagine AI woven into each layer. At the awareness stage, you generate content (blogs, LinkedIn posts) and monitor trends to align with what's on your audience's mind. AI agents can help by mining topics and automating content distribution, so you're consistently present in the right channels. At the interest stage, your personalized outreach campaigns (drafted and scheduled as described above) capture the attention of those encountering your content. At the consideration stage, digital assets like white papers and webinars (and the AI-driven follow-up to those downloads) nurture the lead. Finally, at the decision stage, AI assists with proposal preparation, meeting scheduling, contracting, and even analyzing which touchpoints moved the needle.

AI-enabled workflows not only save time but actually drive better outcomes. In practical terms, an AI-enhanced funnel means fewer leads falling through cracks and more time spent on the human touches that seal deals.

Throughout this funnel, the consultant's role stays central. AI automates research, drafting, and routine outreach—the "grunt work"—freeing you to do what humans do best, which is building relationships, answering nuanced questions, and conveying sincerity. For instance, while an AI agent handles the mechanics of booking meetings and sending reminders, you prepare insightful solutions and empathize with client concerns. This handoff ensures that technology underpins each stage without ever replacing the personal presence that clients need.

Sell Like You Serve

Great consultants don't pitch; they help. AI, when used wisely, enhances your helping. It lets you obtain the right leads, nurture each contact with relevance, and show up consistently. You can deliver insight *before* the sale, sharing ideas and analysis proactively rather than only producing value after a contract is signed.

When you use AI in the spirit of this book—as an empowering copilot, not a black-box replacement—you amplify your own strengths. You'll likely win more business, but more importantly, you'll attract the *right* business. Firms that appreciate your consultative, curious approach will be drawn to your consistent presence and depth of insight. In essence, by selling the way you serve (with clarity, presence, and genuine curiosity), you transform marketing and sales into an extension of client service. As you apply these AI tools and tactics, remember that every email, post, or report is a chance to deepen trust. Use technology to multiply those chances, while you remain the author of the relationship.

ROB BERG

168

CHAPTER TEN

AI-Enhanced Proposals and Contracts

In consulting, the proposal is often the first real expression of your value to a client. While often perceived as a formality you just have to grind through, it's truly your first impression and sets the tone for your entire relationship. A well-crafted proposal showcases your understanding of the client's needs, your credibility, and how you plan to solve their problems. It can be the deal maker or deal breaker in winning a project. As an independent consultant or small firm, you likely pour hours into writing proposals and contracts. This chapter explores how generative AI can help you do it faster and better, while keeping the result human-centered, authentic, and tailored.

We'll look at the strategic value of proposals and how AI can support each step of drafting them—from turning rough notes into polished outlines, to refining tone, structuring deliverables and timelines, and aligning everything with the client's expectations. We'll also dive into how AI assists with pricing strategies (like creating clear value-based or tiered pricing sections) and how it can help with the legal side of consulting—drafting and reviewing contracts, summarizing terms, comparing versions, and flagging risks (always with responsible use in mind). You'll see how to set up automation pipelines to generate proposals or contracts with minimal manual effort and know when a human touch and review are essential to avoid missteps. Throughout, we emphasize that AI is a support tool, not a replacement for you. The best consultants differentiate themselves through authenticity and a coaching mindset—guiding clients and building trust—and courageously leveraging creative new tools like AI. The goal is to let AI handle the grunt work so you can focus on the human aspects of insight, relationship-building, and personalizing your value.

Let's explore how an AI-enhanced approach to proposals and contracts can save you time, increase clarity, and even boost your win rates while ensuring *you* remain in control of the final product.

The Strategic Value of Proposals

Before diving into AI, it's important to remember why proposals matter so much. Many consultants admit they don't love writing proposals; writing can feel tedious or stressful. But the reality is that few clients will hire a consultant without a written proposal; it's the first tangible showcase of your value. It's where you translate all those great conversations and ideas into a concrete plan that the client can evaluate.

A strong proposal isn't just about scope and price; it's a strategic tool that can build trust and set you apart from competitors. When you deliver a proposal, you're telling the client "Here's what I can do for you and *how* I'll do it" in a way that also reflects who you are. It's both a blueprint for the project *and* a marketing document for your services.

What makes a proposal effective? First, it provides clarity and alignment, outlining what will be done, when it will happen, and how success will be measured. This clarity helps prevent misunderstandings later. Second, a great proposal establishes credibility; it demonstrates your professionalism and expertise, giving the client confidence that you understand their problem and can solve it. Third, it serves as a marketing tool, highlighting your unique value proposition and how you differentiate from others. And finally, a well-crafted proposal boosts your chances of winning the business by making it easy for the client to say "yes" with a clear, actionable plan. Proposals that are clear and client-focused increase the chances of winning and lay the foundation for a successful project partnership.

Importantly, your proposal is often the first indication of your working style. It sets the tone for your entire relationship with the client, so make it count! Little things in a proposal can signal a lot. For example, is it organized and easy to follow? Is it written in the client's language or full of jargon? Does it focus on the client's needs, or does it spend pages bragging about your amazing accomplishments? The best proposals keep the focus on the client. As many experienced consultants advise, avoid the common mistake of writing from a "your-firm-first" perspective instead of a "client-first perspective." Remember that it's all about the client and what they need. This aligns with the human-centered philosophy of consulting; show that you understand *them*. Resist the urge to fill your proposal with generic info about your firm and instead tailor it to address the client's specific situation.

Finally, speed and responsiveness are also part of proposal strategy. We live in a fast-paced world. A disproportionate number of wins go to the vendor who responds first to a prospect's request. In consulting, this means being prompt in delivering a quality proposal can literally make the difference between winning or losing the project. A clear, well-structured proposal that arrives quickly (while the competition is still drafting theirs) can impress the client with your responsiveness and preparedness. Of course, speed should not come at the cost of quality or authenticity—and that's exactly where AI can help, by dramatically reducing the time required to produce a polished proposal without sacrificing thoughtfulness.

In short, proposals are *strategic*. They are not boring PDFs to churn out begrudgingly—they're your chance to shine and set the stage for a successful engagement. With that context in mind, let's see how generative AI can support you in crafting proposals that hit all the right points, including client-focus, credibility, clarity, and timeliness while allowing your authentic consulting voice to shine through.

Drafting Proposals with Generative AI

Writing a proposal can sometimes feel like a daunting blank page exercise. But generative AI is like having a tireless writing assistant to help you get started and refine the draft. Rather than replacing you, the AI acts as a speed booster and brainstorming partner in the proposal drafting process. Here are specific ways you can leverage it:

Transforming Notes into an Outline. Most proposals start with raw input, like your notes from client meetings, an RFP document, or a rough idea of an approach. Generative AI (like ChatGPT, Gemini, or Claude) excels at taking disorganized or text-heavy input and structuring it into a coherent outline. For example, you can prompt the AI with a summary of your conversation with the client (their goals, challenges, and the services you discussed) and ask it to draft a proposal outline. In seconds, the AI can produce a structured outline with sections like Introduction, Client Needs, Proposed Solution, Deliverables, Timeline, and Pricing all filled in with bullet points gleaned from your notes.

This approach saves you from staring at a blank page and wondering where to begin. It's like having a consultant colleague quickly sketch a first draft structure for you. ChatGPT is best used as an assistant to streamline the writing process, helping brainstorm ideas and outline the structure rather than writing the entire proposal word-for-word without guidance. In other words, you feed it the raw materials and directions, and it gives you a head start with a logical flow.

Real-world consultants have reported that using AI in this way can drastically cut down the initial drafting time. Instead of spending half a day organizing thoughts, you might get a solid outline in a few minutes which you can then build upon. By accelerating this step, you not only save time but can respond to the client faster, thus gaining a likely competitive advantage.

When generating an outline, ensure your prompt is specific. For instance, *"Here are notes from my call with Client X: [bullet list of notes]. Please draft a consulting proposal outline covering the client's objectives, our proposed approach, deliverables, timeline, and pricing. Use a professional but clear tone."* The more details you provide (e.g., client industry, specific goals, any constraints mentioned), the more tailored the outline will be. You can always refine by saying *"focus more on X"* or *"please add a section about Y"* if the outline misses something important.

Improving Tone and Clarity. Every consultant has a unique style, and every client has a unique culture. One client might appreciate a very formal, buttoned-up proposal, while another might respond better to a friendly, conversational tone. Generative AI is extremely useful for adjusting the tone, language, and clarity of your proposal draft. If you've ever written a proposal and worried whether it sounded too stiff or jargon-y, or whether your value proposition was coming across clearly, AI can help you revise the wording.

For example, you could take a draft of your executive summary and prompt the AI with, *"Rewrite the following paragraph in a more friendly, client-centric tone (8th grade reading level) while preserving the meaning..."* and then paste your text. In moments, it will output a version that might be simpler, more direct, or more upbeat in tone, depending on your request. On the flip side, if you feel your text is too casual, you can ask for a more formal and professional tone. The AI can mimic many styles, from enthusiastic marketing-speak to sober business formal, so you can fine-tune the voice until it resonates with how you want to come across to this client.

Clarity is equally important. Proposals loaded with consultant jargon or run-on sentences can confuse or alienate clients. AI can help simplify complex sentences and ensure key points don't get lost. One best practice is to instruct the AI to keep responses simple and free of jargon. Use simple language your client understands and avoid consulting acronyms. You can even feed sections of your proposal to the AI and ask if any sentence is unclear or too technical and rewrite any such sentences. This can reveal hidden ambiguities or overly dense wording.

Additionally, AI can help maintain consistency of tone throughout the document. Perhaps you collaborated with a team and different sections sound like they were written by different people. You can prompt the AI with something like, *"Make the tone of these sections consistent. Use a warm, confident voice throughout."* This ensures your proposal reads as one cohesive, polished piece, increasing the professional impression.

However, be cautious. Always review the AI's wording to make sure it accurately reflects what you meant. AI might occasionally oversimplify or change nuance. Use it as a helpful editor, but never blindly accept changes if they don't fit your authentic voice or the truth of what you're promising. The goal is to amplify your authenticity, not to produce a cookie-cutter corporate brochure. Remember, if everyone uses the same AI without personal input, proposals could start to sound eerily similar. As such, be sure to augment your prompts with personal examples, insights, and anecdotes, so AI can help you polish them rather than replace them.

Structuring Deliverables and Timelines. Clients pay close attention to what you will deliver and when. These sections (e.g., Deliverables, Timeline, and Milestones) can make or break a proposal's appeal. AI can support you in structuring these pieces in a clear, visually compelling way. For instance, if you list a bunch of activities and dates in rough form, you could ask the AI to organize them into a structured timeline or table.

Say you have a list of deliverables and their target delivery dates. You might prompt, *"Turn the following list of deliverables and dates into a table with columns for Deliverable, Description, and Delivery Date."* The AI can produce a nicely formatted table or at least a well-aligned list that you can then put into your document or slides. This reduces the need to mess with formatting and ensures you haven't missed an item.

When it comes to timeline narratives, AI can assist in describing your project phases clearly. You can prompt something like, *"Help me describe the project plan. We will have five phases, including Discovery, Analysis, Solution Design, Implementation, and Deployment. Draft 1-2 sentences for each phase explaining its purpose and duration in a client-friendly way."* If you supply the specifics (e.g., *"Phase 1, 'Discovery,' will take place over two weeks during which we will interview stakeholders and review documents"*), the AI will turn that into a smooth narrative paragraph or bullet points. This ensures your timeline tells a story of progress which reassures the client that you have a thoughtful approach.

Generative AI can even help identify if something is missing. Suppose you listed deliverables but forgot a section on project management or risks. If you feed the outline or draft to the AI and ask, *"Are there any important sections or details I might have omitted for a consulting proposal?"* it might respond with a suggestion (e.g., *"Consider adding a section on key risks and mitigation, to show you've thought about possible challenges"*). In fact, including a "Risks and Mitigation Strategies" section in proposals is a known best practice, as it shows you're proactive and thorough. AI, having been trained on countless documents, can remind you of such best practices when properly prompted.

Using AI in this way helps ensure the structure of your proposal is complete and logically organized. It's like having an ever-present checklist in your head, drawn from the wisdom of perhaps dozens of successful proposals. Do you have a clear scope section? Did you define acceptance criteria for deliverables? Is project governance explained? AI might prompt you about these if you ask it to review your draft outline for best practices.

Aligning with Client Expectations. One of the philosophical themes of this book is being human-centered and client-focused, and proposals are where that philosophy becomes concrete. The best proposals are tailored to the client's specific context and expectations. AI can assist in this tailoring process in a few clever ways.

First, generative AI can analyze client-provided materials (like an RFP, or even the client's email or website) and help you mirror the language and priorities the client uses. For example, you might upload the client's RFP in which their goals are listed and ask the AI, *"Based on the client's stated goals in the attached RFP, what are the top three priorities we should emphasize in our proposal?"* The AI can parse their language and respond with something like, *"The client is most concerned with increasing conversion rates, improving internal process efficiency, and achieving these within a three-month timeframe."* Those might be things you already know, but seeing it articulated helps ensure your proposal explicitly addresses each one. You can then double-check each major section of your proposal to ensure you're aligning with those expectations. For example, you can ensure that your "Solution" section clearly shows how it increases conversion and your timeline supports how the improvement unfolds within the three-month timeframe specified by the client.

Second, AI can help adjust the style of your proposal to fit the client's culture. If the client is a buttoned-up financial institution, you might want a more formal tone and a very structured format. If the client is a tech startup, maybe a more conversational tone with a short and snappy executive summary is better. You can feed samples of the client's communications (like a portion of the RFP or their "about us" page) to the AI and say, *"Analyze the tone and formality of the following text. Then, adjust my proposal text to match that style."* This is an advanced use, but quite powerful, as it helps you speak the client's language. If the client says "increase sales" rather than "drive revenue growth," you'd want to mirror their word choice for resonance. Using a "client-first perspective" in proposals extends even to word choice and emphasis.

A concrete example of alignment might be ensuring the value metrics you highlight match what the client cares about. Suppose your proposal emphasizes "improving customer satisfaction," but the client in conversation was laser-focused on "reducing costs by 15 percent." AI can catch that potential conflict if you ask it to review with, *"Does this proposal draft emphasize cost reduction in line with the client's request?"* It might answer, *"Cost reduction is mentioned only briefly. The client's main concern was cost, so consider quantifying how your solution saves money."* This kind of feedback can be invaluable in realigning your draft before the client ever sees it.

In summary, AI can act like a coach in the background, nudging you to be more client-centric. It can transform verbose consultant-centric text into crisp statements of client benefit. It can highlight where your draft might accidentally drift away from what the client asked for. By using these AI capabilities, you ensure the final proposal feels like it was written *for that client*, not a generic template. This differentiation through authenticity—showing you truly listened and customized—is what can set you apart. Clients can tell when a proposal is boilerplate. AI gives you no excuse to deliver a generic proposal because it's never been easier to customize content at scale.

Before we move on, it's worth reinforcing the need to *always* review and refine the AI outputs with your own judgment. AI doesn't know your client personally; it's basing suggestions on patterns. You know the client best. Use AI's suggestions as helpful drafts or sanity checks, but make the final decisions yourself to ensure the proposal remains authentic and realistic. When done right, you collaborating with AI can produce a proposal that is both high-quality and deeply aligned with what the client expects to see.

Communicating Pricing Strategies with AI Support

Pricing is a critical component of proposals. It's where the value you bring is translated into dollars. Many consultants find it tricky to present pricing in a way that's clear and compelling. Should you charge a flat project fee, value-based fees, or offer tiered options? Are hourly fees unavoidable due to unknowns or client preferences or procurement conventions? And once you decide, how do you communicate that pricing to the client without confusion? AI can assist both in strategizing the pricing approach and in generating easy-to-read pricing tables or breakdowns.

Crafting Value-Based Pricing Narratives. One powerful pricing approach for consultants is value-based pricing, which means setting your fee based on the value or ROI you expect to deliver to the client rather than based solely on hours or inputs. If you choose this strategy, you need to clearly articulate *why* the price is justified by the value. This is an area where AI can help by wordsmithing the value narrative and even doing some rough quantification if you provide the data.

For instance, you might prompt, *"Help me explain a value-based price. Our project fee is $50,000, which is based on an expected client ROI of at least $500,000 in savings or new revenue within a year. Draft a concise paragraph to justify this pricing in terms of value delivered."* The AI might come back with a paragraph that frames the fee as a small fraction of the potential upside, reminding the client of the specific benefits (e.g., *"...a 10% increase in conversion rate worth $X in new sales"*) to anchor the price in value. By doing so, you're tying the price to outcomes (not hours), which is usually more persuasive.

In fact, when formulating a value-based model, it helps to define the unique outcomes you deliver and quantify the impact on the client's business, then set prices reflecting that value rather than just your costs. Generative AI can assist by taking your list of outcomes and impact metrics and turning it into a compelling value proposition statement. For example, if your consulting is likely to improve efficiency by 20 percent, the AI can help phrase the price discussion like, *"By implementing these recommendations, the client could save approximately $200,000 annually in operating costs. Our fee of $50,000 is an investment toward unlocking those savings."* This sort of wording explicitly connects dollars saved to dollars invested, and when the former is some multiple of the latter, the likelihood of acceptance increases dramatically (assuming you've made a compelling case in support of the expected return).

AI can also ensure you don't forget to mention intangible value. Perhaps your service also provides intangible benefits like risk reduction or future-proofing the business. You can ask the AI, *"What are some value points to justify a premium consulting fee besides immediate ROI?"* It might suggest points about long-term capability building, competitive advantage, peace of mind, etc., which you could then weave into your pricing rationale.

Presenting Tiered Options Clearly. Another strategy is offering tiered pricing. For example, you might offer a Basic, Standard, and Premium package, each with different scopes and fees. Tiered pricing can be very effective for giving clients choice and capturing different budget levels. The challenge is to define the tiers and present them side by side clearly so the client can compare.

AI can help generate the content for tiered packages once you outline the concept. You might input, *"We want to offer three tiers. Basic (minimal deliverables A, B), Pro (everything in Basic + C, D), and Premium (everything in Pro + E, F + extra support). Draft a comparison of these three packages in bullet points or a table."* The AI will structure it as you ask, possibly creating a neat table with each tier's features and price. For example, it might lay out:

- ***Basic Package.*** *Includes A, B; suitable for small teams or pilot projects.* ***Price.*** *$X.*

- ***Pro Package.*** *Includes A, B, plus C, D for a more comprehensive solution.* ***Price.*** *$Y.*

- ***Premium Package.*** *All features of Pro, plus E, F and dedicated support hours.* ***Price.*** *$Z.*

This saves you time formatting and ensures consistency in how each tier is described. Notice terms like "basic, professional, premium" can be generated by AI or you can specify your own labels. Design tiers by offering increasing levels of service to meet diverse client needs. For example, a sample prompt might be, *"Create a basic package for small businesses, a professional package with broader services, and a premium package with comprehensive support and dedicated account management."* That's asking AI to essentially follow a best practice, with each tier aimed at a certain segment.

When presenting tiered pricing, clarity is king. AI can be used to simplify complex pricing info into tables or bullet lists. It will ensure each feature is only listed in the tiers where it appears, avoiding confusion. Once you have the AI-generated structure, *always* do a sanity check. Make sure the differences are clear and that the middle tier isn't lost (a common pitfall is the "Goldilocks" middle option that might not stand out; you may want AI to help emphasize its value too).

Also, consider asking AI for a recommendation. *"Based on the features above, which tier might most clients choose and why? How can I emphasize the value of the middle tier?"* It might respond with advice to highlight how the Pro package has the best value for most clients or include a small discount when compared to buying features à la carte. You can incorporate those suggestions into your pricing section narrative.

Retainers, Subscriptions, and Other Models. For ongoing consulting engagements, retainer or subscription pricing is common, where a client pays a monthly fee for a certain level of access or support. Communicating retainer terms can benefit from clarity and consistency, which AI can provide in drafting.

If you're offering a retainer, you might say. *"Outline a retainer arrangement for me. The client pays $5,000/month for up to twenty hours of consulting support, including X, Y, Z services, with unused hours not rolling over."* The AI will output a paragraph or bullets to formalize the plan, with something like *"Retainer Plan. For a flat fee of $5,000 per month, Client receives up to twenty hours of consulting services. This includes ... If additional hours are required, ... Unused hours expire at month's end, ensuring dedicated availability. This plan provides consistent support and prioritization of Client's needs."* You can adjust the specifics, but notice how AI ensures all the conditions are mentioned.

One prompt example from a marketing context is, *"Outline a subscription-based pricing model for long-term services."* The AI will suggest offering monthly plans with a set of services, allowing scalability based on needs. Translating that to consulting, you might have bronze/silver/gold monthly plans that vary by how many hours you're available or which services are included. AI can help articulate each plan in a parallel way (similar to tiered packages, but for an ongoing service model).

AI is great at creating clear, logical pricing tables. Many clients appreciate a visual like a table summarizing the fees and what's included. For example, a table with columns for "Option, Description, Timeline, Fee" where each row is a package or phase. Even if you ultimately make it prettier in a document or slide, the heavy lifting of aligning features to fees is done. Note that if you do project-based pricing, it's a good idea to break down costs according to sub-deliverables or work streams in a table. AI can ensure you've listed each deliverable alongside its cost contribution if you provide those details.

AI Insights on Pricing Strategy. Beyond drafting the text, AI can act as a sounding board for your pricing strategy. Not sure if your price is too high or low? You could try a prompt like, *"Considering a consulting project for a mid-size company to achieve X outcome, does a $50,000 fee appear to be market rate, high, or low? Provide reasoning."* The AI might not know your exact market, but it might use general knowledge to say something like *"$50,000 for a three-month project for a mid-size company is within normal range if significant ROI is expected, but ensure you articulate the ROI. Many consultants charge based on value; if ROI is high, $50,000 could be a bargain; if low, it might seem expensive."* While you should take such feedback with a grain of salt, it can surface considerations (like mentioning market standards or ensuring ROI justification) that you might want to address.

If you're torn between pricing strategies, you can ask AI to list pros and cons of each in context. It will likely remind you that hourly fees might be transparent but often concern clients due to the uncertainty they introduce, while fixed fees offer certainty but transfer risk to you, the consultant, who is obligated to deliver for a set price. This can guide your decision and how you explain it in the proposal.

Finally, pricing often involves negotiation. AI can help you prepare for that, too. For example, *"If the client says our price is too high, suggest a few ways to adjust scope or payment terms to reduce cost by 15 percent without devaluing our work."* The AI might propose trading a less critical deliverable, shortening the project, or moving to a phased approach. These ideas can be handy to have in your back pocket when discussing price.

In all cases, maintain honesty and realism. Don't let AI "bluff" on pricing or promise value you can't deliver just to justify a number. Use it to communicate clearly, not to oversell. Authenticity in pricing means being transparent about what the client gets and how it benefits them. AI can assist in polishing that message, but the underlying offer must come from you. When done right, you'll present a pricing section that feels fair, professional, and enticing—giving clients confidence that investing in you is the right choice.

AI Assistance in Legal and Contractual Tasks

After a proposal is accepted, consultants often need to draft a formal contract or Statement of Work (SOW) and navigate the negotiation of terms. Even at the proposal stage, some clients want legal terms included (like IP ownership, confidentiality, etc.). Dealing with contracts can be daunting if you're not a lawyer, and even if you are comfortable with legalese, it's time-consuming to scrutinize every clause. AI can offer substantial help in drafting clauses, summarizing contracts, comparing versions, and identifying potential risks in contracts, effectively acting as a junior legal assistant. However, this is an area where responsible use is paramount. AI is not a lawyer, and you must carefully review its output. For best results, let's look at how to leverage AI for such use without getting into legal hot water.

Drafting and Customizing Contract Clauses

Writing legal clauses from scratch can be painful. Common sections like non-disclosure, indemnification, limitation of liability, and payment terms often have standard language. Generative AI has effectively read thousands of such clauses in its training and can produce decent boilerplate text that you can then modify or have reviewed by a lawyer.

For example, you might prompt, *"Draft a consulting services agreement clause about confidentiality. Neither party will disclose each other's confidential information, exceptions for info already public or required by law, etc."* The AI will output a paragraph that sounds like a typical confidentiality clause. It might read something like, *"Both Consultant and Client agree to keep confidential all non-public information obtained from the other party during the term of this agreement. Neither party shall disclose such information to any third party without prior written consent of the other, except (i) as required by law or legal process, or (ii) information that becomes public through no fault of the receiving party... "* and so on. That's a pretty solid starting draft! You save time compared to googling templates or writing it longhand.

Similarly, you can have AI generate a limitation of liability clause or an IP ownership clause by describing what you need. For instance, *"Provide a clause stating the consultant retains ownership of pre-existing IP but grants the client a use license, and the client will own new deliverables produced on their behalf."* Within seconds, you have a draft clause.

However, there's a catch. AI is far from infallible in legal drafting. It may create clauses that sound confident but have inaccuracies or unenforceable bits, or simply wording that isn't ideal. Large language models often hallucinate in legal contexts, deviating from actual legal principles or facts in their outputs. So you should never use an AI-drafted clause verbatim without review. Think of it as producing a rough first draft that you (and ideally a legal professional, if the contract value is high) will refine.

Another issue is consistency. If you ask AI to draft a clause multiple times, you might get different wording each time. It doesn't inherently know what standard phrasing is for a particular clause in your industry. Inconsistency can be risky; if each contract you issue has slightly different indemnity language because AI worded it differently, you might accidentally give one client more protection than another. It's good practice to standardize your key clauses. You can use AI to help create a standard, and once you have one you like, stick with it. Save the best AI-generated version, get it reviewed by a qualified attorney, and then reuse that language for future contracts rather than re-generating each time.

Summarizing and Analyzing Contracts

Contracts often go through multiple rounds of edits and can become long and convoluted. AI is very handy for summarizing key points of a contract in plain English. For example, if a client sends you a ten-page consulting agreement from *their* legal department (some larger clients insist you sign their paper), you can feed chunks of it to an AI and ask, *"Summarize what obligations this contract places on me as the consultant"* or *"List any clauses that define liability or IP ownership and summarize them."* This can save you from misreading or missing something buried in legal jargon. It acts like a second pair of eyes highlighting sections that prevent you from competing with the client for a year, or stating that any work you produce becomes the client's intellectual property. Armed with that summary, you can decide what you're okay with and what to negotiate.

Another use case is comparing contract versions. Let's say you sent your standard contract to the client, they marked it up with redlines (changes), and now you have to see what's changed. Instead of manually hunting through every edited line, you could give AI the original clause and the revised clause and ask, *"What's the difference between these two versions of the clause, and what are the implications?"* For instance, maybe the client edited your payment term from net fifteen days to net sixty days. AI would quickly spot that and might respond, *"The client changed the payment due period from fifteen days to sixty days, meaning they want a longer time to pay invoices."* Simple example, but you get the idea; it accelerates pinpointing changes.

There are also specialized AI tools emerging for contract comparison. For example, some legal AI can take two versions of a document and output a comparison report. Even without fancy tools, ChatGPT itself can do a fair job if you carefully input the text.

Risk detection is another area, like asking the AI to identify any potentially problematic clauses. For example, *"Review the following contract and flag any clauses that might pose significant risk or are unusual for a consulting agreement."* It might come back with, *"Clause 8 (Indemnification) seems one-sided requiring the consultant to indemnify the client for all losses; this is a high risk to the consultant. Also, there is no termination clause, meaning the consultant might be locked in or have no easy way to exit if needed."* This is golden information for you to know where to focus negotiation. AI's pattern recognition can catch things like *"unlimited liability"* or *"non-compete restrictions"*—things that consultants often should negotiate. It's like having a junior legal associate comb through the text first, so you can then address the issues.

That said, don't over-rely on AI. It might miss nuanced issues. It's pretty good at finding the glaring red flags (like the absence of a cap on liability, or indemnities, etc.), but it might not fully grasp the subtler business implications of a clause. Use it as a safety net, not as your sole risk evaluator.

Responsible Use of AI in Legal Contexts

While AI can dramatically improve efficiency in dealing with contracts (by cutting hours of reading and drafting down to minutes, for example) you must use it responsibly in legal contexts. The following points are a good place to start.

- **Confidentiality.** Be cautious about sending full contract text to a public AI service if the contract contains sensitive information. Either use anonymized versions or a secure, enterprise version of AI that ensures confidentiality. Some consultants will replace real names with placeholders before feeding into ChatGPT (e.g., "ClientCorp" instead of the actual client name) as a minimal step.

- **Review by humans.** Always have a human (someone with a legal background) review AI-generated legal text. In proposals and marketing, a minor AI mistake might just be a typo or a slightly off phrase. In contracts, a minor mistake can result in a big liability. Don't skip the human review. If you're a solo consultant not well-versed in contract law, consider at least a quick consultation with a lawyer for mission-critical clauses or use trusted templates.

- **Understand limitations.** AI doesn't know the law in the way a lawyer does. It's not aware of jurisdiction-specific requirements unless told, and it can't give you legal advice. It might draft something that sounds fine but is legally unenforceable or missing mandatory pieces for your type of agreement. For example, AI might not automatically include a clause required by law in some fields (like data protection clauses if dealing with personal data). You need to know enough to prompt for those or add them later.

- **Avoid over-reliance on AI's memory.** If you have a back-and-forth negotiation with multiple changes, be careful about prompt context. If you fed an earlier version and then later feed a new version, the AI might not "remember" the old context unless in the same session, and even then, context windows have limits. It could hallucinate missing pieces if you ask it to fill gaps (*"Add a clause about X"*), so double-check that it didn't accidentally contradict something else in the contract.

All these cautions aside, it's undeniable that AI is transforming how contracts can be managed. Big law firms and small businesses alike are using AI tools to review documents faster. There are now specialized platforms like LegalOn and Harvey.ai that plug into Word and help redline, comment, and suggest clauses automatically. For example, one AI tool can find risks and points of negotiation, effectively redlining an entire contract in record speed. That shows how far the tech has come. Tasks that took lawyers many hours can be done in seconds for a first pass. As a consultant, you may not invest in enterprise legal AI, but knowing that even general tools like ChatGPT can do a large proportion of what those specialized tools do is empowering.

In practical terms, you might use AI to handle the heavy lifting of comparing documents and drafting terms, and then focus your human effort on the parts that require relationship nuance (like how to negotiate a point with the client in a friendly way, or which risks you are willing to accept for the sake of the partnership).

AI can help you translate legal jargon for your client, too. If a client is not familiar with a clause you're proposing, you could generate a short, plain-language explanation of it to include in an email or conversation. This is part of a coaching mindset where you use AI to foster mutual understanding. For instance, *"Our contract includes a limitation of liability clause. In simple terms, this means we are agreeing neither of us will hold the other responsible for more than $X in damages if something goes wrong. This is a standard protection for both parties."* AI can help draft these explanations which make the negotiation smoother and more human-centered.

In summary, use AI to take the tedium and complexity down a notch. Draft the boilerplate, summarize the dense text, flag the glaring issues. But always overlay your own expertise and ethics. When in doubt, default to a professional legal opinion for critical matters. AI's role here is augmentative, giving you more bandwidth and insight so you can negotiate and finalize contracts that protect both you and your client while maintaining a positive working relationship.

Automating Proposal and Contract Workflows

Up to this point, we've discussed how AI can assist in *writing* and *reviewing* proposals and contracts. Now let's turn to the exciting world of automation, where you connect various tools to create a seamless pipeline. Imagine if a significant portion of your proposal generation or contract creation process happened automatically: fill in a few key inputs, and within minutes you have a complete draft ready for review. This is not a futurist fantasy; it's quite achievable today with a bit of setup using tools like Airtable (a flexible database/spreadsheet), Make.com (a no-code automation platform), and of course AI services like ChatGPT.

We'll explore what an AI-powered proposal automation pipeline can look like, and equally important, when to insist on human review despite automation convenience.

Proposal Automation Pipelines: From Client Info to Draft in Minutes

Let's walk through a scenario. Say you have a standardized process for discovery calls with clients and you take structured notes (client name, industry, key pain points, desired outcomes, services likely needed, etc.). All that information could be stored in a tool like Airtable or a CRM. Using an automation service like Make.com or Zapier, you can set up a workflow that triggers whenever you add a new "Proposal Needed" record. The workflow might do the following:

1. **Fetch client data.** Make.com grabs the client's details from Airtable (like their company profile and the notes from your discovery call).

2. **Generate Draft Content with AI.** It then sends a prompt to an AI (via an API, e.g., OpenAI's API) that says, *"Use the following*

information to draft a consulting proposal: [Insert client background and notes]. Structure it with an introduction addressing the client's needs, a proposed approach, deliverables, timeline, and pricing options. Use a professional, yet warm tone. Highlight how our solution is tailored to [Client Name]."

3. **Create a document.** The AI returns the draft text. The automation flow could then automatically insert that text into a proposal template—for example, populating a Google Doc or a PandaDoc template with the generated content in the right spots.

4. **Generate pricing tables.** If you have pricing data (maybe in Airtable fields, like selected tier or custom price calculations), the automation can merge those in as well. PandaDoc, for instance, allows merging fields into a document. So your carefully thought-out pricing for that client (which might be calculated based on a formula) goes straight into the draft.

5. **Deliver for review.** The pipeline could then email you a link to the generated proposal or even create a PDF draft. Some systems can even send it straight to the client, but that's not recommended without review!

This might sound complex, but a little effort can go a long way. There are multiple YouTube tutorials out there where a user demonstrates how they built a custom proposal template integrated with their CRM, allowing them to generate proposals at the click of a button. The key components typically include a data source (Airtable/CRM), an AI step to fill in the narrative, and a document assembly tool (Google Docs, Word, PandaDoc, etc.) to produce the formatted proposal or contract.

The benefits of such automation are huge for small firms. If you can crank out a first draft of a proposal in, say, fifteen minutes instead of four hours, you could dramatically increase your proposal throughput. Consultants who have embraced this are able to send out more proposals (and thus have more chances to win work) without working crazy hours. Even large firms are looking at proposal automation to increase efficiency and consistency. Many consultants' proposal win rates are below 60 percent, so volume and quality both matter—remember, it's been said that 80 percent of success is just showing up. And AI-assisted automation allows you to show up more frequently.

However, automation is only as good as the inputs and design. Garbage in, garbage out. If your notes are poor or you don't prompt the AI properly, you're likely to get a mediocre proposal. You might also end up with a very templated, boilerplate feel if not careful, which could undercut the authenticity we preach. The trick is to design the templates and prompts in a way that still injects your personality. For example, maybe include a field in Airtable for "Consultant personal insight to highlight" and you write a sentence or two unique to each proposal that the AI must include ("Jane's business is at a turning point similar to a past client we helped; I'm excited to bring that experience to the table."). The automation can merge that sentence in an "About Us/Why Us" section, giving a personal touch.

Tools like PandaDoc also offer native features like content libraries and even some AI assistance. PandaDoc's Q&A site outright asks if ChatGPT can write a business proposal and concludes that yes, it can draft one, but it's not likely to differ from others without that added human touch. PandaDoc integrates with Zapier and Make, meaning you could automatically create a proposal document from a template when triggered. So you might store your static content blocks (like your bio, case studies, standard terms and conditions) in PandaDoc, and use AI to generate the custom parts on the fly.

Another aspect is contract reuse and customization. If you have a master services agreement or SOW template, you can use Airtable to store variations of clauses or variables (like payment terms, jurisdiction, etc.). An automation can pick the right clauses based on project type or client region. For instance, perhaps you have one version of a termination clause for short projects and a different one for long-term engagements; the database could store both and a rule chooses accordingly to insert into the contract draft. AI can then smoothly connect the pieces ("stitching" them with good transitions).

Some no-code services even specialize in this. One example is a solution where Airtable data is turned into professional contracts by connecting a Google Docs template and automating PDF creation. Users have reported saving hours each month on generating documents that are now just a button click away. Essentially, if your proposals or contracts have repeating elements and predictable structures, you can automate a lot.

The Human Checkpoint—Don't Skip It!

With all this automation, it might be tempting to go fully hands-off. But don't. Always include a step where a human (most likely you) reviews the proposal or contract before it goes out. Automation can assemble a document in seconds, but it can also assemble mistakes or awkward phrasing just as fast if something misfires. As such, it's worth reiterating the wise words of process improvement pioneer Michael Hammer: "Automating a mess yields an automated mess."

Think of the automated draft as 80 percent complete. The last 20 percent is a careful read-through to catch things like generic company names or old, outdated (or even fabricated) references. Is the tone consistent? Are all placeholders filled? Is the math in the pricing table correct? Does the proposal reflect what you discussed with the client, or did the AI add an irrelevant service because your description triggered a wrong pattern?

Poorly executed uses of mail merge have resulted in emails and proposals that begin with "Dear [Client Name]" because the placeholder didn't populate (we've all seen them), and the same could happen with AI if you're not vigilant. It's these little things that a human eye must catch, because sending out a proposal with such errors can undermine your credibility, negating any efficiency gains.

Moreover, relationship nuance is something AI won't know. Perhaps in your last call the client's tone shifted about a particular deliverable, e.g., they seemed uneasy. You might want to add a line in the proposal to reassure them or adjust the approach on that point. Your automation wouldn't know that unless you explicitly feed it those notes and prompt it accordingly. So the human review is where you incorporate any final custom tweaks that an automated system could not have anticipated. In short, AI and automation get you most of the way, but your professional judgment and personal touch complete the journey.

A Real-World Use Case

Let's illustrate with a reality-inspired example.

Let's say you spend about four hours writing each proposal from scratch and find it draining, so you procrastinate and send just two proposals per month and win one. With AI automation, you could create an Airtable form where you enter client info after each call, including their specific needs. You could then set up a Make.com scenario that takes those inputs and feeds OpenAI to generate a customized proposal draft, merges it into a Google Docs template (including a standard case study slide and contract T&Cs), and pings you when it's ready, reducing four hours of tedium to thirty minutes of collaboration with AI and automation. In that scenario, you could confidently send two to three times as many proposals each month with far less drudgery, and you'll likely find that clients notice how fast and thorough your proposals are. They're not boilerplate; you'll always check that the AI highlighted the client's specific goals and add personal notes in each cover email. The net result is that you'll save eight to ten hours per week on proposal writing that can be used for more client work or simply more free time. This reflects the ideal, where AI acts as a support that enhances human consulting work, not a replacement.

This example shows tangible benefits like time savings and increased capacity. My own colleagues have estimated 50 to 80 percent time savings on certain tasks using AI, and you're likely to see that play out in your work as well. Those weekly time savings add up significantly over months and years.

The bottom line on automation is that it's a game-changer for operational efficiency, but it must be used thoughtfully. You design the process, you decide the rules and templates, and you verify the output. When done right, automating proposal and contract workflows can increase your throughput without sacrificing quality, and let you spend more time on human-centric activities like client interaction, strategic thinking, and yes, even taking a well-deserved break since you're not stuck at midnight formatting yet another statement of work.

Practical Prompt Templates for Proposals and Contracts

To help you get started, here are some prompt templates you can use with ChatGPT or other generative AI tools. These are phrased in a way to yield useful outputs for proposal and contracting work. Feel free to modify the specifics (in the brackets) for your situation:

- **Draft a proposal outline from notes**

  "You are an expert proposal writer. Outline a consulting proposal based on the following client information: [insert key client goals, challenges, and agreed approach]. Include sections for Introduction, Client Needs, Proposed Solution, Deliverables, Timeline, and Pricing. Make the outline logical and client-focused."

  What it does: The AI will produce a structured outline hitting all the key sections, ensuring you have a starting framework.

- **Improve tone and clarity**

  "Act as a writing editor. Here is a paragraph from my proposal: [insert text]. Rewrite this in a more clear and friendly tone, aimed at a non-technical business client, while preserving the meaning."

  What it does: Gives you a revised paragraph that's easier to understand and more personable, removing jargon or overly stiff language.

- **Generate a timeline or schedule**

  "Create a project timeline in bullet points. The project phases and durations are: [Phase 1—2 weeks—description], [Phase 2—1 month—description], [Phase 3—2 weeks—description]. Present it as a list with dates (assuming project starts January 15, 2026)."

  What it does: Lays out a schedule with projected dates and ensures you have a clear plan to show the client.

- **Create a pricing table**

  "Generate a pricing table for a proposal with the following three package options: Package A: $10,000—basic analysis and report.

191

Package B: $18,000—analysis, report, plus implementation work-shop. Package C: $25,000—everything in B + 3 months follow-up support. Present in a clear table form."

What it does: Returns a nicely formatted table or list of packages, which you can directly use or easily convert into a table in your document.

- **Draft a value justification for pricing**

"Help me write a value justification for a $50,000 project fee. The client stands to gain approximately $500,000 in increased sales if our recommendations succeed. Emphasize ROI and that the fee is an investment for long-term gain."

What it does: Delivers a persuasive paragraph connecting the fee to the benefits, useful in a pricing or proposal cover letter section.

- **Summarize a contract clause**

"Summarize the following contract clause in plain English: [paste clause]."

What it does: Outputs a simpler explanation of a legal clause, which helps you and potentially the client understand it without legal jargon.

- **Compare contract versions**

"Original clause: [text]. Revised clause: [text]. Explain the differences and any risks introduced by the changes."

What it does: Points out exactly how the clause was altered and the implications (e.g., *"The revised clause removes the consultant's liability cap, meaning your financial exposure is higher in the new version."*).

- **Identify risky clauses**

"You are a contract analyst. Review the following contract excerpt and list any clauses that could pose high risk to a consultant. Explain why for each: [paste contract or key sections]."

What it does: AI will enumerate clauses that look one-sided or dangerous (like unlimited indemnity, very broad non-compete, etc.) with a brief explanation, aiding your negotiation prep.

- **Flag missing elements**

  "Here is an outline of a consulting agreement I drafted: [outline or list of clauses]. Are there any important contract provisions that seem to be missing or that I should double-check?"

  What it does: AI might respond with suggestions such as *"It doesn't mention governing law"* or *"There's no termination clause specified,"* giving you a reminder of what to add.

- **Rewrite a clause with conditions**

  "Draft a termination clause that allows either party to terminate for convenience with 30 days' notice, and for cause immediately if a breach isn't cured in 15 days. Make it fair and mutual."

  What it does: Produces a balanced termination clause that you can refine further, saving you the trouble of writing it from scratch.

- **Generate a follow-up/next steps section**

  (This is proposal-related) *"Suggest a brief 'Next Steps' section to end my proposal, encouraging the client to proceed. Tone should be confident and helpful."*

  What it does: Provides a short closing paragraph like *"We are ready to move forward at your pace. Upon approval of this proposal, we will schedule a project kickoff meeting... Please feel free to contact us with any questions..."* which provides a nice closure to the proposal.

Using these prompt templates, you can efficiently tackle different parts of the proposal/contract process. Remember to always review and tweak the AI outputs; think of them as first drafts. Over time, you'll learn how to phrase prompts to get exactly the style of response you want. For instance, adding *"in a friendly tone"* or *"bullet points, not too wordy"* can guide the AI's format. If the output isn't quite right, you can refine the prompt further or just edit manually.

The beauty of prompt templates is that once you find ones that work for you, you can reuse them consistently (even incorporate them into your automation pipeline as we discussed). It's like creating your own "AI playbook" for consulting tasks.

Final Thoughts on Proposals and Contracts

Proposals and contracts might not be the glamorous part of consulting, but they are absolutely essential and reflect your professionalism and value just as much as your final work product. Embracing AI in these areas allows you to spend less time on drudgery and more time on what really matters, which is understanding clients, crafting innovative solutions, and building relationships. We've seen how AI can speed up drafting, ensure your language is clear and client-centered, help articulate pricing in compelling ways, and even act as a guardian when reviewing legal terms.

Throughout all of this, it's vital to maintain the core philosophy of human-centered consulting enhanced by AI. AI gives you the superpowers of efficiency and information, but your human judgment, empathy, and creativity remain at the center. Use AI to differentiate yourself by delivering proposals that are not only fast and polished, but also authentically reflect who you are and are uniquely tailored to each client—something a generic AI output alone could never achieve. Adopt a coaching mindset with these tools; guide the AI like you would guide a junior team member, and let it support you, not take over. And don't be afraid to be courageous and creative in how you use AI. The consultants who thoughtfully integrate these technologies will stand out by delivering more value faster and by being adaptable in a changing landscape.

In the end, a proposal enhanced by AI is still *your* proposal. It carries your insight and expertise, just delivered in a more refined package. A contract reviewed with AI assistance is still *your* agreement, hopefully made safer and clearer through AI-augmented diligence. Clients will appreciate the clarity, responsiveness, and professionalism that result.

So, go ahead and experiment with these approaches in your next proposal or contract draft. You might find that what used to be a dreaded task becomes almost enjoyable. After all, who wouldn't enjoy seeing a polished draft come together in a fraction of the time? That's the power of being an AI-enhanced consultant—working smarter, delivering faster, and never losing sight of the human touch that truly drives success in consulting.

ROB BERG

◆

CHAPTER ELEVEN

Framework and Service Development

C onsultants often differentiate themselves not just by the problems they solve, but by the frameworks and methods they use to define those problems and structure their services. A recognizable framework can turn a complex challenge into a clear model that clients understand. Consider how McKinsey's famous 7-S model helped leaders manage organizational complexity by aligning seven key factors, or how Galbraith's Star Model provides a holistic structure ensuring strategy, structure, processes, people, and rewards all work in harmony. Such frameworks bring clarity and consistency to messy situations, acting as a common language that improves communication among team members and stakeholders. They also signal expertise; having a proprietary model or method gives you an edge that reflects *your* distinct sensibilities and sets you apart from competitors.

AI as a Co-Creator

Today, AI offers a generative partner to help consultants accelerate the creation of these intellectual assets. Importantly, we're not talking about AI replacing your human insight; rather, it's augmenting your creativity and speed. Think of AI as a brainstorming collaborator, not a prescriptive engine for answers. You bring the experience and judgment, and AI brings the prompt-based ideation and iteration. This chapter explores how you can leverage that partnership to amplify your consulting practice, covering how to:

- **Develop proprietary frameworks** (e.g., diagnostic tools, strategic models) that make your insights tangible.

- **Reframe your expertise into reusable service offerings** that clients can easily understand and buy.

- **Prototype new offers quickly and test them iteratively**, so you only build what delivers value.

- **Name and language-brand your offerings** for memorability and distinctiveness.

- **Modularize your services** for flexibility and scalability (think "Lego blocks" instead of one-off projects).

- **Productize your consulting IP** into assets that are teachable, scalable, or even licensable beyond your direct involvement.

Throughout, the tone is conversational and pragmatic. This isn't abstract theory but actionable guidance. And in the spirit of the AI-enhanced consultant's ethos, we'll emphasize a human-first approach—using AI to augment your authenticity and courage, not dilute them. By the end of this chapter, you should feel empowered to treat AI as your co-creator in designing a practice as distinctive as your own values and vision.

From Ideas to Frameworks

Frameworks are shorthand for insight. Whether it's McKinsey's 7-S model or the Business Model Canvas, a good framework turns complex, multifaceted problems into a structured format that's easier to grasp and act on. Frameworks make complexity visible and actionable by breaking big concepts into memorable categories or steps. For example, the Business Model Canvas famously condenses an entire business plan into nine building blocks on a single page, helping entrepreneurs "simplify, validate, and communicate" their vision clearly. Likewise, classic consulting models endure because they bring order to chaos. The 7-S framework maps out multiple interrelated elements of an organization so leaders don't miss critical factors during change, and Galbraith's Star Model ensures that when you tweak one part of a business (say, structure), you also realign the other parts (processes, rewards, etc.) for cohesion. In short, frameworks create clarity, structure, and a shared language. They serve three primary purposes for a consultant:

- **Structure your thinking.** Forcing you to organize insights and not overlook key aspects.

- **Signal your expertise.** Demonstrating that you have a well-thought-out approach (your special "method") rather than just ad hoc ideas.

- **Enable your client to act.** Giving clients an accessible map or model they can remember and follow on their own.

A powerful framework is at once an analytical tool *and* a communication device. When a model truly clicks, it becomes the story your client remembers and retells internally. It's the difference between simply delivering recommendations and equipping the client with a mental model they can use long after your engagement ends. If your framework is clear and resonates, your client can explain it to their colleagues when you're not in the room, spreading your impact. In that way, the framework carries your insight forward, creating traction for change.

How AI helps. In the past, crafting a new framework might take weeks of reflection or a team brainstorming off-site. Now you have AI to accelerate that creative process. AI tools can bring coherence to loosely connected observations and give shape to your intuition. For instance, imagine you've noticed three recurring issues in recent client engagements, such as misaligned priorities, poor handoffs, and unclear metrics. You have a hunch these are related but need a unifying structure. You could prompt your preferred AI chatbot (ChatGPT, Gemini, Claude, etc.) with something like, *"I keep seeing the following problems in mid-sized nonprofits: 1. misaligned priorities, 2. poor handoffs between teams, and 3. unclear success metrics. What is a 3-part framework (with a catchy name) that could tie these issues together into one actionable model?"*

The AI might respond with a suggestion such as a *"3P Model"* focusing on Priorities, Processes, and Performance, with a neat description of how each "P" addresses one of the pain points. Suddenly, you've got a draft framework, a concise model that you can refine and present. You can iterate on it by asking for alternative metaphors or additional elements if needed. It's like having a brainstorming partner on call 24/7.

In practice (for example), you might use the AI chatbot to generate four different versions of a transformation journey model for a client, playing with funnels, cycles, and other shapes to depict the model. Ultimately, your AI-assisted brainstorming might yield a staircase metaphor that best resonates with the client's experience. The framework (e.g., "Stepping Up Success") can then become a core part of your proposal, demonstrating a tailored, visual approach that the client is more likely to embrace than an ad hoc recitation of your capabilities and services. In this instance, we're not asking AI to invent your insight; we're using it to surface and shape it faster.

And always keep this in mind: Authenticity matters. Use AI to generate ideas, but choose a framework that feels true to your perspective. The best frameworks often come from patterns you've genuinely observed, combined with a fresh structure that highlights what *you* find important. Don't just adopt a trendy model for its own sake; adapt it to fit your and your client's reality.

Turning Expertise into Services

Many consultants struggle to articulate the difference between a one-off project deliverable and a structured service offering. You might do great work in individual engagements, but packaging that into a repeatable service is another skill altogether. Yet, mastering this packaging is crucial if you want to move from chasing one project at a time to building a sustainable consulting business. Productizing your expertise into clear offerings has multiple benefits. It makes your services easier to market, easier for clients to understand, and easier to scale. In fact, if you rely too much on highly customized, start-from-scratch work for each client, you can end up reinventing the wheel and maxing yourself out. Excessive customization keeps you trapped in a cycle of trading time for money, and paradoxically, prospects might get *less* clarity on what you'll actually deliver. On the other hand, when you offer well-defined packages, clients immediately know what they're getting, for what price, and how it will help them. Clarity sells.

AI can help structure your expertise into clearly defined offers. Rather than inventing a new service from scratch, start by mining your own past work for patterns. A practical approach is to feed case descriptions or project summaries into an AI chatbot and ask it to analyze them for common threads. For example:

- **Input your project histories.** *"Here are five project descriptions in organizational change I've completed..."* (then list key points of each project).

- **Ask GPT for patterns.** *"Based on these, what are the common challenges addressed and the key steps or components I tend to deliver? What would a standardized service offering look like that leverages these commonalities?"*

The chatbot might come back with insights like, *"It appears you frequently help clients with vision alignment, process redesign, and leadership coaching as phases."* It could then draft an outline of a service offering, perhaps with tiers or phases. You might get a suggestion to structure Bronze/Silver/Gold packages (e.g., Bronze includes a diagnostic and recommendation, Silver, a diagnostic plus implementation support, Gold, all that plus ongoing coaching and monitoring for six months). You, of course, will refine and decide what fits, but the AI accelerates the recognition of a repeatable service pattern.

To illustrate, you might take writeups from past process improvement projects and let the AI highlight recurring elements. You might find that patterns become evident in each prior project, such as discovery, analysis, process redesign, change management, etc., even if you hadn't explicitly labeled them as such before. That provides fodder for a framework that can be reused as a signature process improvement service for clients. Each phase could then be assigned a name, a checklist, and even a dashboard of KPIs. By packaging your consulting IP in this way, you graduate from selling one-off engagements to selling a comprehensive program that clients progress through. The offering becomes clearer to explain and market, and clients appreciate knowing there's a proven roadmap.

Note that when you notice repetition in your work, you're probably looking at an opportunity. If you notice you're solving very similar problems for multiple clients, that's a flashing sign that you have an emerging service at hand. Don't ignore it! AI is great at helping you see these patterns in your work that you might miss when you're busy delivering. If you're solving the same problem repeatedly, there's a good chance you can develop a productized service. Use that insight to formalize and name your approach.

With your patterns identified, you can then refine the offering's scope and levels. AI can even help wordsmith the descriptions for each tier of service, ensuring they appeal to different client needs or budgets. For instance, you might prompt, *"Help me describe a three-tier consulting package for strategic planning: basic, standard, and premium. Detail what's included in each tier and what the value proposition is."* The result won't be client-ready out of the box, but it will give you a solid starting draft for you to finalize.

Remember, the goal of turning expertise into services is to create leverage. Instead of every project being unique, you develop a menu of offerings that address common needs in your niche. This doesn't make your work generic; it makes it *scalable*. You can still customize within a framework, but you (and the client) have a clear structure to start with.

Naming and Language Development

The language you use to describe your services matters. In consulting, as in any business, good naming can enhance memorability, signal distinctiveness, and even increase perceived value. A methodology with a catchy name or a clever acronym often sticks in a client's mind far longer than a bland description. Consider the difference between saying "I have a process to improve team performance" vs. "I'll take your team through our Mastering Collaboration Program." The latter hints at guidance and direction, whereas the former could be anyone's generic process. Naming is about branding your intellectual property.

AI can assist here in multiple ways. You can ask ChatGPT or Claude to brainstorm names for a framework, tool, or service. You can have it suggest alternative phrasings for your deliverables that emphasize outcomes or benefits (turning consultant-speak into client-friendly language). AI is also handy for translating jargon into plain language, or for generating analogies and metaphors that resonate with a particular audience. For example, if you're working with tech startups, GPT might suggest names with a tech flavor; if you're working with nonprofit organizations, maybe more mission-driven metaphors.

But naming is as much about emotion as logic. Great brand names often evoke feeling. A name like "RapidOps" feels fast and efficient (good for an operations improvement framework focused on speed). It signals energy and urgency. In contrast, a name like "Clarity Compass" invokes a sense of guidance and insight, perhaps fitting for a strategic planning tool that helps clients find direction. There's no one-size-fits-all formula, and that's where AI's ability to generate dozens of creative options can really shine. You might prompt, *"Give me ten name ideas for a framework that helps tech companies manage post-merger integration, ideally something metaphorical or visual (e.g., related to construction, navigation, or growth)."*

From there, you could get suggestions ranging from Fusion Blueprint to Synergy Scaffold to Growth Grid. Some might be duds, but some could spark that *aha!* reaction. Each conveys a slightly different vibe, so you might consider floating a few options to your client contacts to see which resonates the most.

Of course, while AI can kickstart your naming process, final decisions need your human touch. Check the cultural connotations, make sure the name isn't inadvertently negative or already in use in your field, and ensure it aligns with your personal brand. AI might suggest "Quantum Leap" for your training program, but if there are already five other consultants using that phrase, it won't set you apart. Use AI for the heavy lifting in the ideation phase, then apply your judgment and maybe even a quick informal poll of your target audience to pick a winner.

Also, consider language beyond the name—the vocabulary you use in your marketing materials and conversations. If your framework has steps or pillars, what do you call them? AI can help brainstorm here too. For example, if you have a four-step process, you might prompt, *"What's a good alliterative list of four terms (starting with the same letter) for the key phases of a change management program? The phases involve getting leaders in synch, assessing gaps, implementing solutions, and getting changes to stick."* You might get back something like, *"Align, Analyze, Act, Anchor,"* which has a nice ring to it. Small touches like that give your content polish and make it more memorable.

Don't be afraid to language-brand your approach. Even if it feels a bit salesy, giving a name to your framework or method is an act of ownership over your IP. It also makes it easier for clients to talk about; well-crafted names are clear and imply a defined process. Use AI to help, but ensure the final tone matches your personality. As always, it should feel authentic when you say it.

Finally, keep in mind that fancy names should never come at the expense of clarity. Clever is good, but clear is better. If you can get both, great. If you must choose, lean on clarity, and maybe add a tagline to bring in the emotion. For instance, you might name a framework "The SCALE Model" (where SCALE is an acronym for the steps), but add a tagline like, "Accelerating Growth with Strategic Clarity" to hint at the benefit. AI can help you play with those combinations, too. It's all part of developing a distinctive voice for your services.

Prototyping and Testing with Speed

One of the biggest advantages of AI in consulting is speed, particularly when it comes to prototyping new offerings. In the past, if you had an idea for a new workshop or service package, you might have spent weeks perfecting a detailed proposal or brochure before ever showing it to a client. By then, you've invested so much time that if the client says they're not interested, it's a big loss. Now, you can quickly create a mock-up, get feedback, and iterate, all in a fraction of the time it used to take. This reduces the risk of pouring energy into an offering that flops.

Think of this as running consulting design sprints. Rather than bet all your chips on one idea, you generate two or three plausible variations of an offering in an afternoon, and test them out informally to see which resonates. AI tools are great for this kind of rapid ideation. For example, you can use ChatGPT to help write a one-page concept note or a faux landing page for each version of your idea. Let's say you're considering a new leadership development program specifically for mid-level managers. You could prompt, *"Draft a one-page brochure for a leadership development program*

204

targeting mid-level managers in tech companies, focusing on communica-tion and team motivation. Emphasize results in three months." In seconds, you'll get a draft. Then apply your human touch, and perhaps create an alternate version targeting senior executives, just to compare angles.

What exactly can you prototype quickly with AI? Quite a lot, it turns out:

- **Service brochures or one-pagers.** Instead of perfecting the actual deliverables, mock up a brochure that describes the service and its benefits. AI can fill in placeholders and even suggest agendas or timelines.

- **Sample proposals or pitch decks.** Use tools like ChatGPT or even presentation generators like Gamma to draft a proposal for the new service.

- **"What-if" scenarios for different niches.** Wondering how your offering would appeal to HR vs. Operations vs. Finance leaders? Change a few parameters in your prompt and regenerate the pitch tailored to each persona. You'll quickly see where the strongest appeal is.

- **Visual mockups.** Even if you're not a designer, tools like Canva (which now have AI features) or PowerPoint with a bit of creativity can help you create a sample graphic or process diagram for your framework. AI image generation is emerging too (though use with caution for professional materials). AI-driven tools like Gamma can create decent slide decks from outlines, so you could have it generate a rough "sales deck" for your offer and see how it feels.

The idea is not to have a finished product, but a tangible draft that you can put in front of people for feedback. Share it with a trusted peer, a mentor, or even a friendly client to get their gut reaction.

To systematically test your prototypes, you might even use simple surveys using SurveyMonkey or Typeform. You can embed your mocked-up description and ask a few key questions like, "Would this interest you? What would you expect to pay? What would make it a must-have?" By gathering data or at least qualitative feedback before you build the full offering, you dramatically reduce the chance of a big swing-and-miss. Systematically testing business ideas can dramatically reduce the risk of failure and increase the likelihood of success. Why not bring that discipline to your consulting practice?

Remember, AI is your ally, not your replacement. It can generate content and options, but you still decide what makes sense to pursue and what feedback to listen to. Always loop back to your human sense of what feels right for your business. The goal is not to create an AI-designed practice that doesn't feel like you; it's to use AI to uncover the best ideas you might otherwise not think of and do it faster.

Creating Modular, Scalable Services

As you refine your offerings, another strategy for scalability is to think in modules rather than monoliths. Consultants historically have sold big, custom projects that deploy legions of onsite staff for months or even years at a time. That can be hard to sell to new clients and hard to repeat unless you clone yourself. By modularizing your services, you create a library of components that can be configured according to client needs and budgets. You essentially turn your consulting services into a set of building blocks, which provides tremendous flexibility.

For example, imagine you typically deliver a project that includes an assessment phase, a training workshop, an implementation roadmap, and a follow-up review. Instead of always packaging all four together, you could make each one a module. So one client might buy the assessment and roadmap but not need the training. Another might want the training as a standalone service. Still another eventually buys the full suite (which is effectively all modules combined). Modular design offers an ability to mix and match, making it easier to tailor scope without reinventing the wheel each time.

Advantages of Modular Design

- **Flexible pricing.** Have a small client with a limited budget? Sell them one or two modules. A large client with big needs? Stack multiple modules into a comprehensive program. This "pick and choose" approach is easier to sell at different budget levels.

- **Adaptable across industries.** Because modules are discrete, you can often swap examples or slight context in a module to fit a different industry, rather than redesigning an entire approach. The core stays the same.

- **Easier to scale with help.** If you bring on associates or subcontractors, you can train them to deliver specific modules. Maybe you have a colleague who's great at running workshops who can handle the workshop module while you do the strategy module. Modules make delegation simpler.

- **Consistency and quality.** With defined modules, your IP gets documented and refined each time. You'll deliver more consistent results and continuously improve each component.

AI can assist in modularizing your offerings by helping you break down a big service into its logical parts. For example, you could prompt, *"Here is an overview of my six-month transformation program [insert program description]. Help me break this into four modules, each with a clear focus, a learning goal, key deliverables, and success metrics."* The AI might identify modules like *"Module 1: Discovery & Diagnosis; Module 2: Strategy Co-Design; Module 3: Implementation Sprint; Module 4: Evaluation & Next Steps"* or something similar, with some creative naming thrown in. It could even suggest a clever acronym or alliteration for the modules if you ask. This gives you a starting blueprint.

From the client's perspective, modularization is very appealing. It feels more customized; they can buy what they need, skip what they don't. It also reduces decision friction; a client might hesitate to sign a $100,000 engagement all at once, but be perfectly willing to start with a $5,000 module as a pilot, then add on others once trust is built. Research in service design backs this up. A modular approach enables flexible configuration of services, essentially letting you compose tailored packages for different needs. It provides insight to both you and the client about all the pieces available, which in turn helps them express what they value most and allows shared decision-making on what to include. In other words, it's collaborative and client-centric.

Modularization isn't just operationally efficient for you, it's also empowering for the client. You're giving them a menu of options. They feel a sense of control in choosing what's most relevant to their priorities, budget, and timeline. And as a consultant, you benefit by not having to create a new recipe from scratch each time, you're recombining your proven ingredients. This also means quality improves over time by iterating your modules, making each stronger with each delivery. Over time, your "module library" becomes a real competitive asset.

When defining modules, be sure each module delivers standalone value. A pitfall is to create arbitrary slices that don't make sense on their own. Clients should be able to select modules independently with no further obligation. If one module is dependent upon another, they're not truly modular; they're just steps in a fixed process. Sometimes sequence matters (you usually shouldn't implement before you diagnose, for example). In that case, consider packaging complementary modules (e.g., discovery and analysis) as one module if they're inseparable. Use your judgment on how to slice things. AI can propose options, but you know your practice best.

Lastly, consider preparing collateral for each module. Here again, AI can help. Once you define a module, have AI draft the description, the objectives, maybe even the workbook or tools associated with it. Over time you'll polish these, but AI can spit out a first draft of a slide deck outline for your workshop module, or a template for a report in your assessment module. It's like populating each "kit" in your toolbox with starter content. That way, when someone buys that module, you're not starting at a blank page; you've got materials ready to tailor.

From Framework to Productized Offer

A major goal for many consultants (at least those who want to scale their impact and perhaps decouple time from money) is to productize their services. A productized consulting offer is one that's packaged in such a way that it can be repeatedly sold and delivered in a consistent manner, potentially without your direct involvement every single time. It's moving from custom service to scalable product, turning your framework or methodology into something teachable, tangible, and even licensable.

Productized services may still involve you or your team delivering something, but it's standardized (for example, a fixed-price ninety-minute advisory call every week with set agenda and materials sold as a subscription). Licensing goes a step further; it's getting paid for others to use your IP or framework without you in the room. For instance, you might license your training curriculum to a corporation's HR department, or license another consulting firm to use your proprietary assessment tool in their market. In both cases, you've detached your earnings from hourly work. This allows you to scale without extra work; you build it once, and it generates income multiple times.

AI can significantly aid the journey toward productization by helping create the assets needed for scaling. Here are a few ways consultants are using AI to productize their IP:

- **Creating self-assessment tools.** If you have a diagnostic framework, you can turn it into an online quiz or questionnaire that clients can use on demand. AI can help you formulate the questions and scoring logic. (For example, using ChatGPT to draft a set of questions for each dimension of your maturity model, and even suggest how high and low scores should be interpreted.)

- **Designing maturity models with scoring rubrics.** Many consulting frameworks adopt maturity models (e.g., Level 1-5 of digital maturity). AI can assist in articulating what each level looks like, what behaviors or metrics define each level, etc. This can become a cornerstone of a product if you package it as an organizational assessment tool, for example.

- **Drafting guides and manuals.** To productize, you often need to document your methodology so others can follow it. AI is great at generating first drafts of things like facilitator guides, training manuals, or step-by-step playbooks. For instance, a prompt that asks AI to *"Draft a facilitator's guide for a half-day workshop teaching the first module of my framework"* will yield a structured outline that you can then refine with your specific content.

- **Marketing and sales collateral.** If you're going to sell a productized service or license, you need clear documentation of what it is and how it benefits your clients. AI can help write the marketing copy, slide decks, and even initial scripts for sales webinars or demo videos.

To truly productize, it's useful to check your framework against a few criteria to ensure it's teachable, testable, and transferable (call them the three T's if you like). In other words:

- **Teachable.** Can someone else (besides you) understand and explain your framework easily? If it only lives in your head or requires your charismatic explanation each time, it's not yet a product. AI can help by writing up plain-language explanations and FAQs for your method, which you can give to others to see if they "get it" without your intervention.

- **Testable (scorable).** Can the value or progress be measured? Products often have metrics, like an assessment that produces a score or a traffic light dashboard. If your IP can generate a score or classification (even if qualitative), it feels more tangible. AI can assist in defining those scoring rubrics or categories.

- **Transferable (reusable).** Can your framework apply across different clients and contexts with minimal adjustment? If it's so tailored that it only fits one client, it's not a strong candidate to productize. Aim for frameworks that address common denominators or universal challenges in your niche.

Productization often requires a mindset shift for us as consultants. We're used to customizing and refining and being in the mix. To productize, you have to standardize and let go a little. It becomes less about "What can *I* deliver uniquely?" and more about "What *system* or *asset* can I create that delivers value repeatedly?" AI can expedite creating that system, but you, the human, need to ensure it's high quality and truly valuable. Also, moving to a product mindset means thinking about marketing and sales in new ways (more like a software or info-product company would). But the payoff is that you can grow without being on the hamster wheel of hourly projects. Instead of recreating solutions from scratch, you can package your expertise into repeatable offers that allow you to scale. When done right, clients get a clear, proven solution and you get to extend your impact (and income) beyond your personal bandwidth.

Embrace the fact that productizing might feel uncomfortable at first. But you can do it gradually. You might productize one piece of your business (like an assessment or a training series) and keep doing custom consulting for other things. That's fine. The key is, with AI's help, you can develop those productized components much faster and more affordably than ever before. You can also update them easily. If your framework evolves, you can quickly generate new materials. In a sense, AI is like your R&D department for developing consulting products.

One final note on licensing. If you have something truly unique and valuable, licensing can become a whole strategy in itself. For example, maybe you've created a methodology for project management in healthcare. You could license it to hospitals or to other independent consultants serving those clients. You provide them with the toolkit and training (again, AI can assist in creating user manuals or e-learning content for licensees). In return, you charge a licensing fee or a royalty. This can generate passive or near-passive income. Just be sure you've protected your IP (trademarks, copyrights where applicable). AI can even draft basic legal agreements, though you'll want a lawyer to finalize them. The exciting part is you might wake up one day to find people implementing your ideas in places you've never even been, and you're getting paid for it. That's when you know your consulting practice has evolved to a new level.

Frame It, Name It, Package It

Your ideas deserve structure, language, and a solid form. In this chapter, we saw how AI can accelerate the path from a raw thought in your mind to a concrete framework, from scattered expertise to a defined service, from a rough draft to a deployed offering. If you want to work smarter and scale faster as a consultant, start treating your ideas like products—assets that can be designed, tested, refined, and scaled—rather than one-off projects.

Let's recap:

- **Frame it.** Identify the patterns and insights in your work and turn them into frameworks and models. Use AI to organize your thoughts and generate potential structures that bring clarity. Remember how frameworks like 7-S or the Business Model Canvas gained traction by making complexity simple and communicable. Your own frameworks can do the same for your clients and become a signature part of your brand.

- **Name it.** Don't let your intellectual property hide behind generic labels. Give your frameworks and services memorable names and compelling language. AI can help brainstorm options, but you'll choose the one that resonates. A good name is sticky and sets you apart; it's the difference between being a commodity and having a recognized method. Whether it's a bold metaphor or a crisp acronym, language is part of the value you deliver. It helps in marketing and it helps clients remember you. Even big firms play this game (e.g., Gartner's Magic Quadrant, The Forrester Wave, etc.), because it works to create intrigue and authority.

- **Package it.** Once you have structured and named your approach, package it into services or products. This means defining what you offer in clear terms, maybe in modules, and considering how it can scale. Prototype new offerings quickly with AI's help, get feedback and refine accordingly. Then, when you know you have a winner, standardize it. Create the playbooks, templates, and materials (again, much of which AI can draft for you). The result could be a multitier service line, an online course, a toolkit, or a licensable methodology. The format can vary, but the key is that your IP is now an asset that isn't bound strictly to your hours. You've increased your reach with the same effort, effectively decoupling time from value.

Above all (broken record alert), remember that AI is your assistant, not your replacement. It can shorten the distance between your insight and its impact by handling the tedium of drafting and organizing, but the vision must come from you. Keep your consulting practice human-centered. Empathy, creativity, and trust are things clients will always value and machines can't replace. Use AI to free up more of your time for those human elements. Use it to give you courage to try new things as you become empowered to ideate and test at low cost, thereby venturing into new territory where others haven't dared yet to go.

In the AI-enhanced consulting future, the consultants who thrive will be those who blend their unique human expertise with AI's superpower of generation. So frame it (distill your wisdom into frameworks). Name it (craft the language that sells your value). Package it (build offerings that can scale that value beyond the room you're in). And let AI expedite each step of that journey. The opportunity is immense; you can design a consulting business that's not only more profitable, but also more *you*, aligned with your values and powered by your best thinking, with an AI boost. That's the vision of the AI-enhanced consultant: human-first, authentic, and unafraid to co-create with technology to chart their own path. Now, it's up to you to make it real. Frame it, name it, package it—and go make your unique impact at scale.

ROB BERG

◆

CHAPTER TWELVE

Research and Thought Leadership

Consulting is a thinking profession. The most effective consultants don't just respond to client needs, they anticipate them. They build reputational capital by contributing to the dialogue in their field, offering insights that shape how others see problems and solutions. This chapter explores how AI empowers consultants to conduct research faster, generate insights more consistently, and establish themselves as credible thought leaders.

In the past, building that kind of thought leadership meant countless hours poring over reports, writing and rewriting articles, and staying on top of every industry development. Now, AI is changing the game. By 2024, roughly three-quarters of companies worldwide were using AI in some capacity, and professional services firms (consulting included) saw one of the fastest adoption rates of all. Top consultancies are pouring resources into AI capabilities. Accenture has already secured billions in AI-related work in just one year, and McKinsey noted that 40 percent of its projects now involve AI or advanced analytics (as of 2024). The message is clear; clients expect their consultants to be AI-savvy. Those who leverage these tools effectively can outthink and outpace the competition. But importantly, AI isn't a replacement for the consultant's judgment and creativity; it's an accelerator. It allows you to gather knowledge faster, test ideas, and share insights at scale while you focus on the uniquely human part of the work (building trust, interpreting nuance, and making wise recommendations). In the balance of this chapter, we'll cover:

- Rapid research strategies

- Drafting white papers and articles

- Gathering and analyzing competitive intelligence

- Content repurposing and audience targeting

Throughout, we'll see how an AI-augmented approach can amplify your authentic voice and expertise (never replacing it), and how it encourages a mindset of experimentation, adaptability, and curiosity in your consulting practice.

Rapid Research Strategies

Speed and relevance matter in research. When a client asks a pointed question, or when you're preparing for a high-stakes conversation, having an informed perspective quickly is invaluable. AI tools like ChatGPT, Claude, and Perplexity have transformed how consultants gather and synthesize information almost overnight. Even the largest firms are leaning in. McKinsey & Company built a GPT-powered assistant named Lilli that can sift through a century's worth of internal knowledge to answer consultants' queries. As of this writing, over 75 percent of McKinsey's employees reportedly use Lilli regularly (around 500,000 prompts per month), and a firm case study found it cut the time spent on certain research tasks like information gathering and synthesis by about 30 percent. Boston Consulting Group has taken a similar approach with an AI chatbot called GENE, which is trained on BCG's best thinking and industry data to serve as a conversational research partner; it's like having a tireless junior analyst on call at all times.

Of course, using AI effectively means learning to ask good questions. When McKinsey first rolled out its AI tools, many consultants experienced "prompt anxiety"—uncertainty about what to even ask these systems. A little practice goes a long way to overcome that. Think of prompting an AI like coaching a junior researcher; if you give clear instructions and context, you'll get surprisingly relevant answers. If you ask something vague, you'll get a vague response. Don't be afraid to experiment with different prompts; you can always refine and try again (the AI never gets tired of your follow-up questions!). And remember to stay curious and humble in the face of what the AI delivers; treat it as a starting point for deeper inquiry, not the final word.

AI Can Help You: *(rapid research tasks)*

- Summarize long documents, reports, or transcripts in seconds

- Synthesize perspectives across multiple sources

- Identify trends, contradictions, or gaps in current thinking

- Suggest follow-up research areas or client-specific implications

Workflow Example. Let's say you need to get smart on a niche topic in an afternoon. Here's one way to do it:

1. Use Perplexity (or a similar AI search tool) to gather five to seven reliable sources on the topic.

2. Ask Claude to identify thematic patterns or key insights across those sources. (For instance, *"What are the common trends and any conflicting points these reports mention?"*)

3. Use ChatGPT to rewrite those findings into a concise industry briefing or internal memo, tailored to your client's context. Using ChatGPT's "Deep Research" capability or its "Agent Mode" delivers astonishingly detailed reports that are comprehensive and well-documented. (As always, check the output for accuracy!)

Prompt Template (for research synthesis):

> *Summarize key takeaways from this report on the future of employee experience in hybrid workplaces. Identify trends, risks, and implications for midsize companies.*

Using a prompt like that, you can feed in a dense report and get a tailored summary of what matters most in plain English in a matter of seconds.

For example, imagine you're preparing for a strategy session with a regional healthcare network. You need the latest insights on healthcare labor demographics and digital health innovations. You can use AI to summarize multiple analyst reports in less than an hour, something that might take a full day if performed manually. This enables you to join your strategy session with up-to-the-minute insights on demographic shifts, workforce availability, and "digital front door" patient engagement trends. Your clients will undoubtedly be impressed with the quality and timeliness of information you have at the ready.

The goal is not to offload thinking to the AI, it's to get to the thinking faster. By removing the grunt work of skimming, compiling, and aggregating sources, AI lets you focus on interpretation and application. It's like having an intern who can speed-read one hundred documents and give you the highlights, while you do the critical analysis. (Of course, you still need to vet the AI's output. Treat it as a fast but occasionally error-prone assistant. Double-check important facts or numbers, just as you would review an analyst's work.) Used wisely, AI ensures that your limited research time is spent understanding the story the data is telling, not just hunting and gathering information.

Drafting White Papers and Articles

Thought leadership is a credibility builder. It shows clients and prospects what you've done and how you think. In consulting, sharing your ideas publicly can differentiate you from the pack. But many consultants struggle to find the time (or the confidence) to write regularly. Staring at a blank page or trying to craft your thoughts into a coherent argument can be intimidating, especially after long days at work. This is where AI shines as a writing partner.

AI Can Help You: *(help with content creation tasks)*

- Convert an outline or a rough transcript into a polished article

- Repurpose long-form writing into bite-sized blog posts or LinkedIn threads

- Offer structural suggestions, compelling headlines, and catchy title options

These capabilities aren't just theoretical. For instance, as mentioned in the Preface of this book, I developed an "alignment guide" and "style guide" to ensure AI-generated output matched my writing style and sensibilities. What used to take a human editor hours now happens in seconds, ensuring every white paper or report (or book) *sounds* consistently like me. (I've really been amazed and how well ChatGPT incorporated words or phrases I use frequently in this book, as well as captured specific values or philosophies I have regarding AI and the consulting profession in general.) Tools like ChatGPT or Grammarly can serve as your personal editor, polishing your prose and catching grammatical slips. And if writer's block is the issue, generative AI can help you get past the blank page. Imagine you have a handful of bullet points or a voice memo of an insight; you can feed that to ChatGPT and ask it to draft a first version of your article. Suddenly, you're not starting from scratch anymore; you're reacting to a draft, which is so much easier than conjuring text out of thin air.

For example, suppose you want to write a piece on a new sustainability regulation but you're not sure how to structure it. You could prompt an AI with, *"Here are five points I want to make about this regulation. Can you draft an article that weaves these together, with an intro and conclusion for an audience of supply chain executives?"* The AI will produce a reasonable draft that you can then refine. As I've continued to use AI for writing in different formats and contexts, the tools I use gain "memory" of my writing style and get better at producing narratives that genuinely capture my thoughts in a way I might have struggled to get down on paper. Seeing my ideas laid out boldly by the AI actually helps me to build confidence in my own voice. You might find a similar boost as the AI-generated text gives you something to agree or disagree with, which clarifies your own perspective.

That said, remember that *your* perspective is the star of the show. AI might suggest a turn of phrase or a neat analogy, but you decide if it truly fits what you want to say. The aim is to amplify your voice, not replace it. If an AI-generated sentence doesn't sound like you or runs counter to your point of view, change it or toss it. Be especially mindful of "AI tells" that structure sentences in a format that screams "AI generated!" One example I found more than fifty instances of in the drafting of this book (that I manually cleaned up) included variations of the phrases *"Not only will this make you a better consultant, it will make you a superstar!"* or *"Using AI not only supercharges your abilities, it sets you miles apart from your competitors."* The "Not only *this* but *that*" structure using grandiose or hyperbolic language is a dead giveaway that AI had a hand in the writing. Leaving in a couple of instances is fine; but fifty or more in a 300-page book is a no-go. Authenticity is key in thought leadership; readers can tell when something feels off. Use the tool to enhance clarity and style, but make sure the ideas and tone still feel like you.

Workflow Example (from idea to publication):

1. **Start with a core insight.** Dictate your idea out loud—maybe a lesson learned from a recent client engagement or a new trend you've been thinking about. (Don't overthink it; just speak as if explaining it to a colleague.)

2. **Transcribe and summarize.** Use a tool like OpenAI's Whisper (or Microsoft Word, which has reasonably good transcription capabilities) to transcribe your spoken thoughts. Then, if needed, ask an AI to summarize or clean up the transcript. You'll end up with a rough text capturing your insight.

3. **Draft with AI assistance.** Paste that text into ChatGPT or Claude and prompt it to generate a first draft in a chosen format. For example, *"Turn this concept into a 1,000-word article. Make sure to include an introduction, subheadings for each major point, and a conclusion with a call-to-action."*

4. **Repurpose into multiple formats.** While you're at it, why not get more mileage? Ask the AI to also produce a 300-word blog post version of the article and a fifty-word teaser you could post on social media to draw people in. (You can even request a tweet thread or a slide outline—whatever formats matter to your audience.)

5. **Review and refine.** Now you have a few pieces of content. Your job is to review each one critically. Edit the language to sound like you. Add or correct details based on your expertise. Maybe sprinkle in a quick example or story only you would know. This is where your judgment and experience polish the AI draft into a credible piece.

6. **Publish and share.** Post the article on LinkedIn or your firm's blog. Publish the shorter blog post on your website or as a LinkedIn

article. Use the teaser (or those tweets) to draw attention to the full piece. You've now broadcast your insight across multiple channels with relatively little extra effort.

Consider following a process like this to turn a routine project debrief into a multi-part article series. Upon finishing an engagement, you might record a few voice notes about key lessons and let AI help draft the articles. You can then refine them as appropriate and publish on LinkedIn as a series, perhaps over several weeks or months. The result? You're likely to amp up the number of impressions (due to repetition) and might even invite inquires for speaking engagements or new project work. By showing how you think about the project work and extending your reach, you can boost your credibility and attract opportunities instead of constantly chasing them.

Consistency is key. Establish a writing rhythm. If you can put out one valuable insight per month, you'll quickly build a library that reflects your thinking and attracts your ideal clients. Consistency matters more than perfection. (Yes, even in thought leadership, showing up is most of the battle.) AI can help by making the content creation process less burdensome, but you still need to provide that spark of insight and hit "publish." Over time, as you see the positive reactions— a client mentioning your article in a meeting, or a prospect citing your post as the reason they called you—you'll build the confidence to share even more. It's a virtuous cycle as AI frees up a bit of time and removes some friction, which lets you be more courageous and creative with your content. And the more you share, the more you solidify your reputation as a leader in your space.

(On a practical note, always respect confidentiality. When using real client scenarios in your writing, anonymize and change details, and avoid feeding any confidential data into public AI tools. Use common sense. Stick to insights and stories you're free to share, and when in doubt, err on the side of caution or use an enterprise-secure AI platform.)

Competitive Intelligence and Market Awareness

Staying current with competitors' moves and overall market dynamics is essential for strategic positioning. Knowing what others in your space are up to, including new offerings, hires, partnerships, and content pushes, helps you advise clients on how to stay one step ahead (and helps you position your own firm, too). Traditionally, competitive intelligence could be fairly labor-intensive as it involves tracking news, subscribing to newsletters, manually reading earnings reports, and maybe even mystery shopping. AI makes it dramatically easier to gather and analyze this kind of intelligence in real time.

Imagine waking up, grabbing your coffee, and diving into the morning brief your AI assistant prepared. It might read something like this:

> *Good morning! Here's a recap of what happened in your industry in the last 24 hours: Competitor A announced a partnership with a major cloud provider, likely to bolster their AI capabilities. Competitor B posted three new job openings for cybersecurity consultants, hinting at an upcoming push in that area. Meanwhile, a Reddit thread trending in r/InsuranceTech has several customers complaining about slow claims processing at Competitor C, which could be a weakness to highlight in conversations with prospects.*

This sounds like a report an entire research team would compile, but AI can do much of it automatically. By setting up the right AI-driven monitors, you can have a radar that's always scanning the horizon for you.

AI Can Help You: *(competitive intel tasks)*

- Monitor competitors' product launches, press releases, hiring trends, or partnership announcements
- Summarize earnings calls, quarterly financial reports, or investor presentations to catch strategic signals
- Track content themes in competitors' blogs, webinars, or social media posts (what topics are they pushing?)
- Analyze sentiment in forums, reviews, or social media discussions to see how the market feels about different players

Tools to Explore: There are many tools and techniques to automate competitive monitoring:

- Feedly AI, Perplexity, or Google Alerts for curated news feeds on specific companies or topics

- ChatGPT Agent Mode (or Claude with web access) to quickly comb for and summarize online articles, reports, or forum threads

- Job board aggregators and hiring trend analyzers (to see who your competitors are hiring and infer their focus areas)

- API-based monitors for LinkedIn and X to analyze social engagement and trending topics related to competitors or industry hashtags

Prompt Template (for competitor scan):

Here's a prompt you can put to use right away to create genuine value in seconds:

> *Scan the top three competitors in the Insurtech consulting space. Summarize any recent strategic moves (new services, partnerships, etc.), notable hiring trends, and the key themes from their latest blog posts or LinkedIn updates. Then suggest what these moves might mean for new entrants in this market.*

An AI given that prompt (with maybe some links or data provided) could return a concise briefing such as,

> *Competitor X launched a subscription-based consulting package for small insurers and has been posting about agile transformation, indicating a focus on flexible delivery models. Competitor Y made a partnership with a data analytics startup and their recent hires include two senior AI specialists, suggesting they're investing in AI-driven insurance solutions. Competitor Z's blog emphasizes cybersecurity in insurance, perhaps to differentiate themselves. For a new entrant, this implies that a niche focus (like cybersecurity or AI for insurance) could help stand out, and subscription pricing might be becoming a client expectation.*

In practice, you might set up a simple system where an AI agent monitors key review sites and forums for competitors in your industry or niche. If it happens to flag a surge of negative comments about a competitor's new software tool (for example, on Reddit or Gartner Peer Insights), you could quickly publish a blog post addressing that very pain point (in a general way, not naming the competitor). Prospective clients searching for solutions who read your post would genuinely appreciate that you acknowledged the issue and offered advice. This kind of agile response to intel can turn a competitor's headache into your advantage.

Keep in mind that competitive intelligence is not about copying what others do, it's about positioning yourself (or your client) wisely. AI helps you understand the terrain so you can choose where to differentiate. The goal isn't to one-up a competitor on everything; it's to find the gaps or the underserved areas where you can stand out. Because AI can cast a wide net and connect dots in disparate data, it often reveals strategic patterns that would be easy to overlook. It's then on you, the consultant, to decide what to do with that information. Maybe it's doubling down on a niche your competitors are neglecting, or reassuring a client that their rivals' cool new initiative isn't as game-changing as it sounds (because the AI helped you see the potential pitfalls). Also, keep in mind that more data isn't always better. It's easy to drown in information. The consultant's skill (augmented by AI) lies in discerning which intel matters. Use these tools to stay informed, but always filter the findings through your experience and your client's unique context. In competitive strategy, your human judgment is the secret sauce that turns data into decisive action.

Building Your Personal and Firm Brand

Thought leadership doesn't exist in a vacuum. It's part of a broader strategy to build your reputation and your firm's brand. By consistently sharing valuable insights, you:

- **Establish credibility.** People begin to see you as the "go-to" person for certain topics.

- **Attract the right clients.** Your content acts like a magnet for organizations that value the perspective you're putting out there.

- **Influence the industry conversation.** Over time, your ideas can shape how others frame the problems and solutions in your field.

AI can assist in this branding and thought leadership journey in several ways:

- **Content calendaring.** Not sure what to write about next? AI can analyze trending topics in your industry or even your own past content performance to suggest ideas. For example, if you're a supply chain consultant and suddenly "nearshoring" or "resilient logistics" spikes in the news, an AI might flag that and even propose to write about how mid-sized manufacturers can nearshore effectively.

- **Persona-targeted messaging.** Good thought leaders tailor their message to their audience. AI can help you reframe your insights for different personas. Suppose you have a great idea about implementing AI in finance operations. How you present that idea to a CFO vs. a CTO will differ. You could prompt an AI to *"Take my five key points about AI in finance and rewrite them as a memo from a CFO's perspective,"* and do the same for a tech audience. The result is two versions of your message, each speaking the language and priorities of the reader. This can significantly increase your content's resonance because it shows you understand your audience.

- **Content repurposing.** We touched on this earlier, but it's worth emphasizing: AI makes it much easier to turn one piece of content into many. A single webinar can become an email newsletter, a series of tweets, a LinkedIn article, and a few short videos if you have the right help. There are AI tools that, given a video or podcast, will automatically generate transcripts, pull quotable snippets, and draft social media posts. Even without specialized tools, you can do this manually by feeding your webinar transcript into your favorite chatbot and prompting it to *"Extract three pull quotes, five tweet-sized insights, and a two-paragraph summary for a blog."* You'll get a bunch of raw material to edit and post across platforms. This kind of smart repurposing ensures you meet your audience wherever they like to consume content, without reinventing the wheel each time.

Let's walk through a quick workflow example that leverages AI for multi-channel branding:

1. **Record or reuse content.** Take your latest podcast episode, webinar, or a keynote speech. (If you don't have one, you could even record yourself talking for five minutes about a topic you care about.) Feed that audio/video into an AI transcription tool to get the text.

2. **Ask for repurposed pieces.** Plug the transcript into GPT and prompt it with something like: *"Give me five insightful pull quotes from this content, three engaging tweets (each under 280 characters) highlighting key points, and a short-form blog post (200-300 words) summarizing the main idea."*

3. **Review for tone and add your touch.** You'll receive a bunch of content drafts. Review each one. Make sure the tone sounds like you (or your firm's brand voice). This is where you add a bit of personal commentary. Maybe the AI pulled a quote, but you want to make it a bit punchier, or you want to add context to the blog summary. Do that. The AI gives you the clay, but you mold it to fit your style.

4. **Schedule and publish across platforms.** Now you have a mini "campaign" derived from one piece of original content. You can schedule the tweets across a week, post the short blog on your site or LinkedIn, and use the pull quotes as graphics or text posts on Instagram or LinkedIn. By spreading it out, you reinforce a consistent message without sounding repetitive; different formats will reach different people.

Using this method, a consultant can maintain a steady drumbeat of thought leadership even during busy client periods. You might call this "content stretching," since you take a one-hour webinar and use AI to stretch it into a month's worth of posts. It'll keep you visible and relevant, with minimal incremental effort.

If you're not yet ready to publish under your own name or feel you don't have the writing chops, you can use AI to create ghostwriting drafts that others (or even professional editors) can refine and release. Many solo consultants quietly do this; they'll generate a solid article with AI, then have a content editor polish it and maybe publish it on the company blog or industry trade journal. It's a way to build a portfolio of thought leadership without personally agonizing over every paragraph. Over time, as you see those ghostwritten pieces succeed, you'll likely gain the confidence to put your own byline on more content. The important thing is that the ideas *do* originate from you; the AI and editors are just helping to express them. And as always, ensure the final product aligns with your authentic views and voice. You don't want to wake up and not recognize an article attributed to you!

A final note on brand building: Thought leadership is fundamentally about generosity and perspective. You're giving away some of your knowledge for free to help others or to provoke thought. This builds trust. When people see that you consistently provide value without a hard sell, they begin to trust that engaging with you (as a client, partner, or hire) will be worthwhile. AI can help you deliver that value more frequently and finely tuned to your audience, but the spirit of generosity has to come from you. Share insights, answer common questions, address uncertainties in your industry. In a way, you're coaching the market by helping people understand issues and possibilities. That coaching mindset (one of the superpowers of great consultants) translates beautifully into content. And with AI taking care of some of the grunt work, you can focus on delivering that human touch in your writing or videos—the empathy, stories, and encouragement to think differently.

From Insight to Influence

AI allows you to do the mechanics of thought leadership faster and better. It helps you:

- **Research faster.** No more weeks lost down information rabbit holes; an AI can fetch answers and summarize sources in a snap.

- **Write better.** You have a built-in editor and brainstorm buddy, ready to help refine your arguments and wording.

- **Publish more consistently.** With automation and repurposing, you can maintain a regular cadence of content without burning out.

227

That covers the volume and efficiency side of the equation. But thought leadership isn't just about pumping out content. It's about making an impact with that content. It's about:

- **Making people think differently** by offering a perspective or insight that reframes the reader's understanding of a problem.

- **Making the invisible visible** by surfacing an issue or opportunity others haven't noticed yet. (This is where your unique experiences and observations are gold.)

- **Showing up with generosity and perspective** by creating value for your audience, whether or not it immediately benefits you, and doing so with your own personal flair.

When AI accelerates the mechanics, it frees you to focus more on the message. This is a subtle but important shift. Instead of spending five hours on formatting, fixing grammar, or compiling data for an article, maybe now you spend one hour on those things (with AI's help) and you gain four extra hours to think deeply about *what* you want to say. What contrarian view might be worth exploring? What story from your client work really illustrates the point? How can you challenge your readers a bit, or inspire them? These are the human elements that AI can't supply. They come from your creativity, courage, and empathy.

Don't be surprised if working with AI actually pushes you to clarify your own thinking. It's common, for example, to ask an AI to draft something and then realize it's not quite what you meant, which forces you to articulate more clearly what you *do* mean. It's a like when a junior consultant drafts a slide for you. If it's slightly off, you then have to clarify your vision to get it right. In this way, AI can act as a catalyst for sharpening your ideas.

It's also worth noting that the consulting industry is figuring this out right now. There's no one-size template for success. Even the big firms are experimenting, learning, and sometimes stumbling. The firms that are doing well are the ones fostering a culture of experimentation and learning. Consider approaching AI from top-down and bottom-up, where leadership encourages AI use, but also watches for clever uses employees come up with on their own. Often, the most creative applications arise from consultants in the trenches playing around with the tools in unexpected ways. The takeaway for you is to approach AI with curiosity and a willingness to try new things. Maybe you'll prompt ChatGPT to do something no one in your firm has tried yet. Maybe you'll invent a new way to visualize data using an AI image generator, or use a chatbot in a workshop to gather live input from participants. Who knows? The field is wide open, and as a result, you can contribute to shaping how AI is used in consulting.

All this experimentation does take a bit of courage. It's easy to stick to familiar methods, but as we've discussed throughout this book, courageous consulting (shameless plug for my last book) is about forging your own path and trying novel approaches when they could add value. AI is a perfect arena for this kind of courage. There's no real risk in trying an AI tool 'on the quiet' to see what it can do. Worst case, you wasted twenty minutes on a bad output; best case, you discover a hack that saves you and your colleagues dozens of hours. When you find something that works, share it! Become a coach and mentor for others in your orbit who are just starting their AI journey. You'll solidify your own learning in the process and enhance your reputation as an innovator—a nice bonus for your personal brand.

Finally, keep the human-centered mindset front and center. The irony of AI in consulting is that the more we automate and augment, the more our distinctly human qualities stand out. Relationship-building, trust, empathy, and ethical judgment become even *more* important when technology is everywhere, because they're what set you apart from a clever machine. Use the efficiency gains from AI to invest in those human aspects. If AI helps you draft a report in half the time, use those precious, saved hours to have coffee with a client stakeholder or to coach a junior team member. Those are investments AI can't make for you, but they pay huge dividends in your effectiveness as a consultant and leader.

In short, AI can take you from *insight* to *influence* faster than ever before. But the influence part comes from touching minds and hearts with your insights. AI will give you the speed; you must bring the heart and courage to say something that matters.

Be Known for Your Thinking

Being a thought leader isn't about having perfect ideas or writing like Hemingway. It's about consistency, relevance, and a clear point of view. With today's tools, every consultant, including you, has the ability to contribute meaningfully to the conversations that matter in your domain.

Here's how AI enables you, in a nutshell:

- **Faster synthesis of complex topics.** What used to take days of research can often be done in hours, without sacrificing depth.

- **Easier production of publishable content.** You don't need to be a world-class writer to put out a solid article or post. AI can help organize your thoughts and tighten up the language.

- **Stronger awareness of what competitors and the industry are saying and doing.** You can stay in the loop effortlessly, which means your commentary will be timely and well-informed.

Done right, AI-enhanced research and content creation doesn't just keep you personally informed; it helps you become the person others turn to for insight. You move from consumer of information to a shaper of information.

So, as you embark on your AI-augmented thought leadership journey, start by asking yourself a few questions:

- What am I genuinely curious about right now? (Chances are, if you find a topic fascinating or important, others in your field will too.)

- What do my best clients keep asking me? (Their frequently asked questions are goldmines for content ideas. If one client needs to know, many others probably do as well.)

- Where is the industry conversation missing something that I could add? (Is there a perspective no one's talking about? An insight from adjacent industries that could be applied? A myth you can bust with your experience?)

The answers to these questions will point you toward the topics and ideas worth sharing. Then use AI to explore those ideas (research, gather data, test assumptions), draft your thoughts (outline, write, refine), and share them widely (publish, repurpose, amplify).

Every consultant has insights worth publishing. AI just removes much of the friction that used to hold us back. The playing field is more level than ever. You don't need a ghostwriting team or a PR agency to be heard; you need consistency, authenticity, and the willingness to let technology assist you.

No one becomes a recognized thought leader overnight. But with each insightful LinkedIn post, each well-researched white paper, each useful answer you publish on a forum or in an article, you build your credibility brick by brick. AI can help you lay those bricks faster and more firmly by ensuring your contributions are frequent, timely, and polished. The foundation is your expertise and values; that's the core that never changes.

In the end, being an AI-enhanced consultant means you're leveraging every tool to amplify your impact. You're become known for delivering for clients *and* for your thinking—the way you illuminate problems and envision solutions. That is a powerful reputation to have. It attracts opportunities, fuels career growth, and, perhaps most importantly, it means you're genuinely advancing your field.

So don't hold back. You have something unique to say, and now you have new ways to say it. The consultants who embrace this, who marry their human intuition and expertise with AI's capabilities, are keeping up with the times while defining the future of the profession. It's time to be bold, experiment, and let your ideas shine. The world is waiting to hear your voice.

ROB BERG

◆
_____

CHAPTER THIRTEEN

Project Management and Client Engagement

Project management is the connective tissue of consulting. It's how ideas become outcomes and how client confidence is built or lost. As a solo practitioner or leader of a consulting team, you know that managing projects is about more than just ticking boxes; it's about keeping everyone aligned, informed, and confidently pressing ahead. In this chapter, we explore how AI tools are reshaping the way consultants plan and deliver engagements, assess risk proactively, and facilitate better client decisions.

Done right, AI becomes a behind-the-scenes teammate, handling the tedious coordination and analysis so you can focus on what truly matters—the strategic thinking, relationship-building, and creative problem-solving at the core of your work. Accordingly, we'll look at four key areas in this chapter: status reporting automation, risk assessment tools, AI-supported decision matrices, and client engagement.

Each section will show practical ways to use AI as a support tool. Instead of providing one-size-fits-all rules, these examples invite you to imagine and craft your own approach. Think of it as inspiration for augmenting your practice by using AI to amplify your best qualities, while you remain firmly in the driver's seat.

Status Reporting Automation

Timely, transparent updates are critical for maintaining client trust. Yet compiling status reports often becomes a time-consuming chore. If you're like many consultants, you might spend hours each week pulling together project notes, meeting recaps, and updates from team members, only to deliver a report that may go unread. All that effort is time you could have spent solving problems or connecting with stakeholders.

AI offers a better approach. Instead of acting as a passive scribe, an AI can serve as your real-time synthesis engine. It can continuously digest the stream of project information and condense it into the highlights that matter to your client. In effect, it handles the mundane aspects of reporting so you can focus on more important work.

Here are a few ways AI can help generate and streamline your status updates:

- **Extract key updates.** Rather than manually hunting for what changed this week in meeting notes, emails, and task trackers, you can ask an AI assistant to scan through transcripts or project management tools and pull out the relevant progress.

- **Summarize milestones, roadblocks, and next steps.** A good status report tells the story of progress. AI can be prompted to summarize which milestones have been reached, what blockers (if any) came up, and what's coming next, giving your client a clear narrative of the project's trajectory.

- **Format reports in the client's preferred style.** Whether your client expects a slide deck, a brief email, or an update in a shared workspace, AI can help format the output accordingly. For example, you might use a standard template and have the AI fill in the blanks, or generate a draft that you convert to a polished PDF. You spend less time on formatting and more on substance.

In practice, imagine it's Thursday evening and you need to send a weekly update Friday morning. You gather the week's meeting transcripts and task updates, then feed them into your favorite generative AI tool. You request a concise summary that includes progress made, issues noted, and upcoming steps. Within moments, you have a draft status report. You review it (adding a bit of your personal tone or emphasis where needed) and then drop it into the client's template or email format. What used to take you a couple of frantic hours now takes maybe fifteen minutes. Your prompt may look something like this:

> *Based on the meeting notes and project tracker updates*
> *provided below, draft a brief status report for the client.*
> *Include bullet points under (a) accomplishments this*
> *week, (b) issues/risks, (c) next steps, and (d) any key de-*
> *cisions made. Use a professional but friendly tone.*

In this instance, your AI assistant will produce a structured summary that concisely communicates all important information. You might be surprised at the details it catches—perhaps a decision from Tuesday's call that you overlooked—ensuring nothing important falls through the cracks.

Even as a one-person consultancy, you can now deliver the kind of polished, consistent reporting that used to require a whole project support team. Better yet, you've freed up time to invest in more important things, like checking in with the client about how they feel the project is going, or tackling a stubborn issue that needs your expert attention. The AI handles the routine assembly of information, but *you* still provide the insights and human touch that make the report meaningful.

You can also tailor AI-generated reports to different stakeholder needs without much extra work. From one set of source notes, the AI could help you produce a detailed technical update for the IT team, a high-level summary for the executive sponsor, and perhaps an easy-to-read dashboard update for end users. All versions stay aligned because they draw from the same data, but each is framed appropriately for its audience. That means less manual rewriting for you, and better communication for everyone.

With routine updates handled, you can turn your attention to heading off problems before they escalate. Next, let's look at how AI can help you anticipate and manage project risks.

Risk Assessment Tools

Every project carries uncertainty. Identifying and planning for risks is a hallmark of professional delivery. However, in a small practice, risk management often stays informal or reactive; you might not formally log a risk until its potential impact has materialized. AI can change that by bringing more structure and foresight to your process, essentially acting as a second pair of eyes to spot trouble early.

Think of AI as a diligent assistant that combs through project information and flags potential issues. It won't replace your judgment, as you still decide which risks are real and how to handle them, but it can ensure you aren't blindsided by something you overlooked. Here are some ways AI can enhance your risk management capabilities:

- **Exposing hidden risks from project data.** Feed your project charter, plans, or meeting notes into an AI and ask, *"What potential risks do you see?"* The AI might notice an extremely tight timeline (schedule risk), ambiguous requirements (scope risk), or a stakeholder who hasn't been engaged (engagement risk). These are hints that might validate your gut feelings or reveal new concerns.

- **Categorizing and prioritizing risks.** AI can help sort risks by their likelihood and impact if you ask it to. This transforms a vague list of worries into a structured risk register. For example, it could label a possible budget overrun as "high likelihood, high impact" (red alert) vs. a minor training hiccup as "low impact" (keep an eye on it). By quantifying risks in this way, you and your client can focus attention where it matters most.

- **Suggesting mitigation strategies.** Based on patterns learned from countless scenarios (or from historical data you provide), an AI can propose mitigation ideas for each risk. If "vendor delay" is flagged, the AI might suggest, *"Schedule buffer time for vendor deliverables, and have a backup vendor in mind."* You'll refine these suggestions with your expertise, but having a starting point saves time.

- **Simulating "what-if" scenarios.** To help a client truly understand a risk, you might ask the AI to draft a brief scenario of what happens if that risk comes true. For instance, you could ask, *"What happens if the schedule slips by a month?"* and get a short narrative of the ripple effects. Making a risk tangible like that can motivate stakeholders to take it seriously.

A prompt to an AI might sound like, *"Here are our project goals and notes from kickoff. Identify five key risks that could threaten success. For each, estimate if it's high, medium, or low likelihood, describe the potential impact, and suggest one mitigation step."* This can yield a solid first draft of a risk log. Often, the AI will articulate some risks that you sensed but hadn't yet put into words. That helps you communicate concerns early, rather than ignoring your intuition. Of course, you'll review and edit the list; maybe the AI flagged something that isn't a big deal in your context, or you need to adjust the wording. But it's much easier to refine a draft list than to start from scratch.

In practical terms, you might use ChatGPT or a similar tool to generate and update your risk list, then track those risks in a shared document or spreadsheet that the client can see. Some consultants set up an interactive risk dashboard (for example, in Notion or Airtable) that is updated as conditions change. If you get new information, like a stakeholder raising a concern in an email, you can ask the AI to incorporate it into the risk log with an updated status or mitigation.

A final thought on risk assessment is to invite the client into the risk management process. Share the initial AI-generated risk list with them and ask if they ring true, or if they have anything to add from their perspective. When clients collaborate in defining risks and strategies, they're more invested in the outcome. It turns risk management into a team effort rather than a solo exercise, building trust because the client sees you're not hiding or minimizing potential issues.

By getting a handle on risks early and maintaining open dialogue about them, you position yourself as a steady, prepared partner for your client. Next, you'll often find that big decisions need to be made during a project, and that's another area where AI can lend a helping hand.

AI-Supported Decision Matrices

Consultants often help clients navigate tough choices, such as which technology to implement, or which initiative to prioritize. Decision matrices are classic tools for bringing clarity to such dilemmas. They provide a structured rationale for choices, but building these matrices manually can be tedious, and bias can creep in if you're not careful.

AI can be a powerful ally in making decision analysis both faster and more objective. As a solo consultant, you might not have a whole team to crunch numbers and explore scenarios, but with AI, you can approximate that capacity on demand. Here's how AI can assist in structured decision-making:

- **Brainstorming criteria and options.** You can ask an AI to suggest what factors to consider for a given decision. For example, if a client is picking a software vendor, the AI might propose criteria like cost, integration complexity, user-friendliness, support quality, and scalability. It might even surface less obvious factors (e.g., data privacy compliance or vendor reputation) that you and the client can then discuss. This way, you start the decision process with a well-rounded view of what matters, rather than overlooking something important.

- **Number-crunching and scenario testing.** Once you have your options, criteria, and weights set, AI can handle the tedium of scoring and recalculating. Instead of manually reworking a spreadsheet repeatedly, you can plug your assumptions into an AI and ask things like, *"How would the rankings change if we doubled the weight of customer support?"* or *"Recalculate if Option B's cost is 20% higher."* The AI will quickly spit out the new totals or rankings. This helps identify if a decision outcome is sensitive to certain assumptions. (If small changes in weights flip the top choice, that's a flag to discuss with the client.)

- **Unbiased pros/cons summaries.** Provide the AI with facts about each option, and it can list the key advantages and drawbacks of each in an impartial way. For a client deciding between entering Market A or Market B, for example, the AI could lay out pros and cons for each based on available data. You'll still verify the points, but it gives a starting draft that is less likely to be slanted by personal bias.

The goal isn't to have the AI make the decision, but to equip you and your client with clearer insight. In practice, you might go through a dialogue with a tool like ChatGPT. Feed it context and raw data, get structured outputs (criteria lists, scoring tables, etc.), refine those with your own insights, and iterate. You remain the guide interpreting the analysis, while the AI handles the tedious computations and exploration of alternatives.

For example, suppose your client needs to choose a new CRM system out of four candidates. First, you list the options for the AI and explain the context (e.g., *"a mid-sized nonprofit, budget-conscious, needs ease of use"*). Next, you ask the AI to suggest decision criteria. It returns a solid list, including cost, ease of integration, time to implement, staffing needs, user experience, support, scalability, potential ROI, and strategic fit. You and the client agree on the top five criteria and assign weights reflecting what matters most. Then you gather whatever data you have on each option, estimate scores for each criterion (with AI's help where needed), and let the AI calculate the weighted totals to rank the options. Perhaps Option A comes out slightly ahead of Options B and C. You then explore a what-if scenario with, *"What if scalability matters twice as much?"* The AI adjusts the weights and shows that under those conditions, a different option would win. This insight reveals *why* the ranking is what it is and how sensitive the outcome is to certain assumptions. Your prompt may look something like this:

> *Create a decision matrix comparing three product launch options. Criteria are cost, time to implement, potential ROI, and strategic fit. Assign reasonable weights and scores for each, then rank the options from best to worst based on the total score.*

A prompt like that will lead the AI to output a possible weighting and scoring. You'll likely get a nicely formatted breakdown showing each option's score on each criterion and the overall totals. Treat this output as advisory, not gospel; it simply provides a framework that you and your client can adjust. Maybe the AI's initial ranking holds true after you verify the inputs, or maybe you realize there are qualitative factors the matrix doesn't capture (for instance, one option might pose cultural challenges that aren't reflected in a numeric score). The real value is the structured conversation it enables. Instead of a vague debate, you and the client have a tangible model to play with by evaluating alternative scenarios such as, *"What if we place more importance on long-term ROI and less on upfront cost?"* and so on.

A bit of upfront structure can prevent a lot of second-guessing later. Clients are more likely to stick with a decision they helped shape with clear rationale. AI's role is to reduce the grunt work and potential bias in building those decision models. It tirelessly handles calculations and what-ifs, so you can focus on interpreting results and guiding the discussion.

Now that we've covered keeping projects on track and making informed decisions, let's examine how to keep the client closely engaged throughout the journey.

Client Engagement

All the efficiency gains from AI won't mean much if the client feels disconnected or in the dark. Project management isn't only about timelines and tasks, it's also about communication, trust, and making the client feel like a partner in the journey. Fortunately, AI can help here as well, not by replacing your personal touch, but by making it easier to keep clients informed and involved.

Think about the best client relationships. There are no surprises, no last-minute panics, and a sense that you're working *with* the client, not just for them. AI tools can reinforce this by enabling real-time information sharing and quick follow-ups with minimal effort. Here are a few ways to enhance engagement and transparency:

- **Live project dashboards.** Even if you're a solo consultant, you can set up a simple project dashboard that updates automatically. For example, you might use a shared online sheet or project management board that an AI (via integrations) updates as tasks are completed or timelines change. The client could have access to this live dashboard to check progress anytime. Instead of sending periodic "percentage complete" emails, the information is continuously visible. This level of transparency builds trust; the client sees you have nothing to hide and can literally watch the work progressing. Just be sure to present the data in an accessible way (highlight key milestones, use plain language) so the client isn't overwhelmed by raw details.

- **Instant meeting summaries and action items.** After a client meeting, it's a best practice to send a recap. AI makes this nearly frictionless. If you record the meeting or take detailed notes, you can feed that text to an AI and get a summary of decisions and next steps. Within minutes, you have a draft email outlining what was discussed and what the next steps are. You edit it for tone and accuracy, then send it out. Clients often appreciate the prompt follow-up as it shows you're on top of things. It also ensures everyone has a written reference, reducing miscommunication. (How many times have projects derailed because people had different recollections of a meeting?) By using AI to speed up documentation, you keep momentum going and demonstrate reliability.

- **Monitoring client feedback.** You can also use AI to spot patterns in the client's feedback or questions. For example, if the AI notices that "timeline uncertainty" comes up repeatedly in meeting notes, you know to address scheduling concerns proactively. This helps you catch subtle signals and respond before small worries become big issues.

In using AI for client communication, one golden rule remains: always add the human touch. The tools can generate content and updates, but you should review and tailor them to sound like *you*. Clients will notice if they suddenly get an email that reads as if a robot wrote it. Instead, use AI to do the hard part (compiling info, drafting text), then infuse the final product with your personality and empathy. For example, an AI-drafted summary might be technically correct but a bit dry. To counter this, you could add a personal opening line thanking everyone for their time, or a note of enthusiasm about progress made. These small edits reassure the client that you are fully present and invested, and not simply running them through an automated process.

By consistently keeping the client in the loop and documenting decisions, you create a sense of security and partnership. The client sees progress and issues as they emerge, rather than hearing about them later. This openness often encourages the client to be more forthcoming as well. In a transparent relationship, they trust you enough to speak up early about their own concerns, because they see you as a collaborator who shares both good and bad news.

One note of caution: don't let the convenience of AI lead to complacency. Always double-check what the AI produces and be mindful of confidentiality. These tools should help you serve the client better, never compromise their trust.

Done right, incorporating AI into client engagement lets you deliver clarity and responsiveness that clients will remember. You're effectively removing friction from communication. Status updates, follow-ups, and analyses happen faster, so you spend less time writing emails and more time having real conversations. Meanwhile, the client feels included every step of the way. They never have to wonder what's going on, because you consistently keep them informed. That consistency builds a deep trust.

With your project running transparently and your client highly engaged, you've set the stage for success. To close this chapter, let's recap how these AI-enhanced practices in project management and engagement contribute to you being a truly effective, trusted consultant.

Clarity, Confidence, and Coordination

When project management is done well, it becomes almost invisible as things just flow. Your client sees progress, feels informed, and grows to trust your process. Used wisely, AI helps you reach this state by reducing friction and revealing insights early. It's not about using flashy tools for their own sake; it's about giving you and your client clarity, confidence, and coordination.

In this chapter, we've seen how AI can help you to:

- **Communicate progress clearly.** Routine updates and dashboards keep everyone on the same page.

- **Spot risks before they explode.** Potential problems are flagged and addressed early.

- **Make well-structured decisions.** Choices are backed by data and a clear rationale.

- **Keep clients engaged and confident.** Transparency and quick communication foster trust.

No matter the tactic, AI is there to amplify your capabilities, not replace the human touch. By automating grunt work and speeding up analysis, it frees you to focus on empathy, creativity, and big-picture thinking—the things that build true trust.

As you integrate AI into your work, remember there's no one-size-fits-all solution. Use the ideas here as inspiration and adapt them to fit your own practice. Ultimately, being an AI-enhanced consultant means partnering with technology to deliver smarter, smoother outcomes while remaining unmistakably human in how you lead and connect with clients.

ROB BERG

244

CHAPTER FOURTEEN

◆
_____

Deliverables, Reports, and Presentations

C onsulting insights are just part of the job. Expressing them clearly, credibly, and persuasively is what makes them count. The most brilliant analysis means little if it isn't communicated effectively. In this chapter, we explore how AI tools can support the production of high-quality deliverables. From drafting and refining reports, to visualizing data, to assembling client-ready slide decks, consultants now have access to generative AI (tools like OpenAI's ChatGPT, Google's Gemini, or Anthropic's Claude) that accelerate polish without sacrificing professionalism. More importantly, these tools allow you to focus on what matters most, such as insight, narrative flow, and strategic relevance. It's no surprise that adoption has been swift, with hundreds of millions of daily users. The key is using these technologies to augment your consulting craft, not replace it.

Why Deliverables Still Matter

Even in an era of real-time conversations and co-creation, deliverables remain an anchor of consulting engagements. They capture agreements, guide implementation, and justify decisions. They also serve as tangible artifacts of the consultant's thought process long after the meetings end. In many clients' eyes, if it's not documented, it didn't happen. A well-crafted report or presentation solidifies your recommendations and becomes a reference that circulates within the client organization.

AI doesn't replace your thinking; it helps it to have impact. By handling grunt work (like formatting or initial drafts), AI lets you invest more energy in the message itself. That's critical, because a deliverable is more than a file; it's a vehicle for influence. Below are core deliverable types in consulting, all of which can be enhanced with the right AI support:

- **Executive summaries** are concise high-level overviews for senior stakeholders.

- **Diagnostic reports** include detailed findings from assessments or analyses.

- **Implementation plans** are roadmaps that translate strategy into action.

- **Strategic frameworks** use models or visuals to simplify complex ideas (e.g., 2x2 matrices, maturity models).

- **Slide decks and workshop materials** include presentation slides, handouts, and interactive content for meetings.

- **Visual models and templates** include diagrams, infographics, and tools clients can reuse.

In addition to your findings, your clients want to know what they *mean*. AI can help with both, but only if you stay in control of the message.

Slide decks shouldn't just tell a story about your findings and recommendations; they should enable your client to retell that story to others. AI can't create that level of utility on its own, but it gives you the space to get there faster. In other words, by automating the busywork, you gain time to craft a narrative is easily digested and truly sticks.

Drafting and Refining Written Deliverables

Reports and memos remain consulting staples. Whether you're submitting a one-page recommendation or a forty-page market entry analysis, clarity is critical. Drafting these documents is often time-consuming, but this is where AI shines. Generative AI can serve as a tireless writing assistant, helping you go from rough ideas to polished prose more efficiently. Your AI assistant can help you to:

- **Turn bullet points or rough notes into full paragraphs.** Have a page of jotted insights? Feed them to an AI like ChatGPT or Claude and get a first draft narrative in return.

- **Reframe content for clarity, conciseness, or tone.** You can ask the AI to simplify technical jargon for a lay audience, make your wording more concise, or adjust the tone (more formal, more upbeat, etc.) depending on the stakeholder.

- **Translate findings into executive-ready summaries.** AI is excellent at condensing detailed analysis into an executive summary that highlights key points and implications.

In practice, imagine you have raw notes or an outline for a report section. You might use a workflow like this:

1. **Input your key points.** Dictate or paste bullet-point notes into the AI chat (e.g., *"Key findings from our supply chain analysis: ... "*). Tools like ChatGPT or Claude excel at taking raw text input.

2. **Specify the audience and tone.** Tell the AI who this write-up is for (*"This is for the client's COO; keep it high-level and strategic"*) and the style you want (*"professional and consultative tone, avoid technical jargon"*).

3. **Ask for a draft and refinements.** Prompt the AI to generate a full paragraph or section. If the result is off-base, ask it to revise (*"Make it more concise"* or *"Emphasize the cost implications more"*). You can even request multiple options for phrasing to choose what fits best.

You ask an AI to *"Take the following notes and write a concise, executive-friendly summary. Emphasize implications, not just observations."*

From the response to that initial prompt, you can iteratively refine the output. The AI becomes a first-pass editor that never gets tired of reworking a sentence.

To provide another example, suppose you had twelve pages of typewritten notes filled with observations from a half-day strategy session. You could feed those notes into ChatGPT and quickly get back a well-structured four-page draft report. You could then run a secondary prompt to distill that into a one-page executive summary and an infographic or bullet list for a slide. The final product, refined with your personal touches, could be delivered on the same day as the session. Your clients will no doubt be impressed with how clearly the report summarizes important takeaways from your workshop and the speed with which you were able to deliver it.

Your AI Tool Stack

For drafting and refining text, consultants typically combine a few AI-powered tools:

- **Content generation.** ChatGPT, Gemini, or Claude to draft paragraphs and sections from outlines or notes.

- **Editing and tone refinement.** Grammarly or Wordtune to polish wording, fix grammar, and ensure the tone/reading level fits the audience.

- **Integrated writing environments.** Notion AI or Microsoft's Copilot in Word to get AI suggestions directly within your document as you write.

Using AI in writing can dramatically compress the drafting timeline. I completed a 3,500-word white paper for my firm in less than two hours, from concept to final text—a process that easily consumed two to three times as much time drafting and editing prior to using AI. These tools can level up anyone's game. The takeaway is that by offloading some of the heavy lifting to AI, you free yourself to focus on high-value aspects, sharpening the logic, injecting your expertise, and tailoring the message to your target audience's situation.

From Raw Data to Strategic Insight

Consultants often work with client spreadsheets, databases, dashboards, and operational reports. The ability to spot trends in data and translate them into meaningful or actionable insights is a core differentiator of great consultants, and one where AI offers massive leverage. A lot of analysis work is repetitive or tedious, and automating it creates more space for strategic thinking. Here's how AI can help you turn raw data into insights and visualizations:

- **Clean and prepare data.** For example, quickly remove duplicates, fill in missing values, or standardize inconsistent formats. It's remarkable that analysts can spend up to 80 percent of their time just cleaning data and only 20 percent analyzing it. AI data-cleaning tools can dramatically cut this prep time.

- **Crunch numbers and find patterns.** AI can perform trend analyses, calculate variances, or identify statistical outliers in seconds. Instead of manually running pivot tables for every hypothesis, you can ask an AI assistant to explore the data and flag anything notable.

- **Highlight anomalies or red flags.** If there are unusual spikes or out-of-range values buried in a report, AI can flag them for you to investigate further, ensuring you don't miss something important.

- **Generate charts or tables for presentation.** Many AI tools can produce basic charts given data. You might, for instance, paste a table of sales figures into an AI chat and request a bar chart of sales by region, or use a specialized tool to create a visualization, saving you the hassle of fine-tuning chart settings in Excel.

In practice, let's say you're handed five departmental budget Excel files, each with different column structures and messy labeling. Rather than spending hours manually reconciling them, you use ChatGPT's Advanced Data Analysis tool. In one session, it ingests all the files, normalizes the data formats, flags discrepancies (like two departments using different names for the same expense category), and even generates a PivotTable-style summary with narrative interpretation. In short order, you have not only clean consolidated data but also a plain English explanation of key spending trends, all generated while you focused on other tasks.

The grunt work of data analysis can now be sped up drastically. In addition to numbers, AI can tackle textual data at scale too. The latest LLMs allow for large context windows (greater than 1,000,000 tokens, or roughly 750,000 words), which means you can feed hundreds of pages of material into it for analysis. What a human analyst might read in five hours, your chatbot can digest in under a minute. Users have dropped in things like entire annual reports or lengthy policy documents and then asked the AI for summaries, risk assessments, or Q&A across the content. In consulting terms, imagine uploading a massive compilation of client research that included market research PDFs, prior strategy decks, and interview transcripts, and asking the AI to pull out common themes or contradictory points. You'd get a head start on analysis that could have taken a non-AI-enhanced team days to sift through.

By automating data crunching and initial interpretation, you can devote your energy to the truly human part of the job, which involves interpreting why those trends matter and advising the client on the implications. The AI might tell you *what* the data says; it's your job to explain *so what?* This human-in-the-loop approach ensures that insights are both data-driven and contextually savvy.

AI for Presentations and Frameworks

Crafting a compelling visual narrative is another area where AI can elevate your work. Consultants frequently need to convert complex analyses and recommendations into presentations, slides, and visual frameworks that resonate with clients. This is a creative task that's part art and part science—and AI won't replace the art. But it can significantly speed up the science (the production of slides and graphics) and even enhance creativity by offering fresh ideas.

Slide Deck Generation. Generative AI has reached the point where it can create draft presentation slides from a simple prompt or outline. Specialized tools like Gamma or Tome allow you to input a rough outline or a text description of a topic, and they will produce a multi-slide deck complete with suggested layouts and imagery. Likewise, major software providers are embedding AI in familiar tools. Microsoft's PowerPoint Copilot, for example, can take a prompt like, *"Create a 5-slide presentation summarizing our Q3 market analysis findings"* and generate a starter deck in moments. It will populate slides with placeholder text (drawn from your prompt or any source document you provide) and even add AI-selected images or icons. The slides won't be client-ready out of the gate, but they give you a solid first draft to build on. Instead of staring at a blank PowerPoint, you instantly have something to critique and edit. Even speaker notes can be auto-generated to help you explain each slide's content.

Visual Design Assistance. Beyond content generation, AI can assist with design polish. Tools like PowerPoint's Designer or Canva's Magic Design use AI to recommend layouts, color schemes, and graphics once they see your content. This means you can move more quickly from a rough-looking draft to a slick, professional-looking deck. If you have brand guidelines, many of these AI features can incorporate those (for example, keeping to the client's color palette or using their approved font). The result is that you spend less time tinkering with alignment and more time refining your message. And for those truly custom visuals, like a unique illustration to visualize a concept, image-generating models like ChatGPT 4o or Gemini Flash 2.5 Image (affectionately known as "nano-banana" after its original internal code name) can create original artwork or icons in a flash. Many consultants are now using AI to generate custom visuals (like a metaphorical image of a "north star" for a vision slide) instead of relying solely on generic stock photos.

Frameworks and Diagrams. Consider the classic consulting presentation artifacts, like 2x2 matrices, pyramids, flow charts, and ecosystem maps. AI can support the creation of these as well. For instance, you could ask ChatGPT to suggest a structure for a framework (*"What are some good ideas for quadrant categories in a 2x2 matrix used to evaluate digital readiness vs. organizational agility?"*) and it will happily brainstorm options. It will likely propose axes or categories you hadn't considered, sparking your thinking. For diagramming, tools like Lucidchart now have AI assistants where you can describe a process flow in words and get a first-pass diagram as a starting point. The AI won't know your client's business like you do, but it can handle the draft work of drawing boxes and arrows once you decide the conceptual structure. This is especially handy for consultants who aren't graphic design whizzes; you get a presentable framework illustration without painstakingly aligning shapes in PowerPoint for an hour.

Workshop and Training Materials. Presentations aren't only for final readouts; consultants also facilitate workshops and interactive sessions. AI can help here too. Need an icebreaker tailored to a client's situation? Ask an AI to generate a creative question or scenario. Preparing a role-play exercise for a leadership training workshop? You can have AI draft a realistic scenario or dialogue as a starting point. Some facilitators use AI-driven transcription tools during workshops (e.g., to capture notes on a virtual whiteboard) and then have a model summarize the discussion in real-time to share with participants. For example, you could record a brainstorming session and by the session's end have an AI-curated list of the top five themes that emerged, ready to present back to the group. This real-time synthesis ensures nothing important gets overlooked, and it demonstrates responsiveness, all while you remain focused on engaging the room, not frantically taking notes.

Always bear in mind that AI gives you draft outputs, but you provide the nuance that only a human can deliver. In visual storytelling, that means you decide which chart truly supports the insight, which metaphor will resonate with this client, and which details to highlight or omit. An AI can generate a slide about, say, a market entry strategy, but you must ensure it's the right slide with the right emphasis. Always align visuals to the story you need to tell; don't let flashy AI-generated graphics distract from the core narrative. Used thoughtfully, these tools speed up the process and spark ideas, but the storytelling remains your domain.

Structuring and Packaging Deliverables

How content is organized often matters as much as what's in it. A poorly structured report or disjointed deck can dilute the power of your insights. Here, AI can act as a smart sounding board for structure. It's your virtual colleague who will review your outline and suggest improvements in real time. Specifically, AI can support structuring by:

- **Suggesting logical sections and flow.** If you provide a brief description of a project, an AI can propose an outline (e.g.,

"Background, Findings, Recommendations, Next Steps") and even sub-points that you might include in each section. This is especially useful when you're staring at a blank page unsure how to start organizing your thoughts, or, likewise, find yourself staring at a page full of steam-of-consciousness ramblings and have no idea how to assemble them into a coherent presentation arc.

- **Flagging redundancies or gaps.** By reviewing a draft, the AI might point out if you've essentially made the same point in two different sections, or if a key question seems unanswered. For instance, after being prompted to review a report, the AI may respond with, *"Sections three and five seem to overlap—consider merging them"* or *"You introduce a risk in the introduction but don't address it later. Would you like me to add a section on mitigation?"*

- **Offering alternative headings or titles.** Perhaps your section title "Market Analysis" is too bland. You could prompt the AI to suggest something a bit more impactful and have it respond with, *"What the Market is Telling Us."* Similarly, for an overall report title, you can ask for options that sound more engaging or concise, then pick or adjust the best one.

- **Reformatting content for different deliverables.** You can even use AI to re-package the same content in different ways. For example, after writing a long-form report, you might ask the AI to produce a one-page abstract or a set of slide titles and bullets derived from it. This ensures consistency in messaging across different formats, i.e., the report, the slide deck, and the email summary, without manually rewriting each from scratch.

In practice, you might prompt with, *"Here's the draft structure of my report: 1. Background, 2. Analysis, 3. Recommendation, 4. Implementation. Suggest a more compelling structure or sequence if you think something would flow better. Also, flag any jargon and propose clearer wording for those parts."*

In response, an AI might suggest breaking the analysis into two sections (e.g., separating quantitative findings from qualitative insights) and point out that a term like "synergy realization" could be replaced with plainer language for clarity. This kind of feedback can significantly sharpen your deliverable before it ever reaches your client.

Integrating AI into your packaging of content can improve the efficiency and effectiveness of your deliverables. Clients receive information in a format that's easier to digest and aligned to their priorities, which reflects well on you as the consultant who understands their issues. Don't hesitate to leverage these tools to ensure your reports and decks are structured in the clearest, most persuasive way possible.

Integrating AI into the Production Workflow

The key to sustained value from AI is integration. This elevates AI use from a one-off trick to a seamless part of how you produce deliverables project after project. In practice, this means developing repeatable workflows and habits around AI usage. Consider these tips for building a system rather than just ad hoc usage:

- **Save prompts as templates for reuse.** If you find a prompt that works well (e.g., a great prompt for summarizing meeting notes into takeaways, or for generating risk/mitigation tables), save it. Build your own prompt library that includes tested prompts for common consulting tasks. This way, you're not starting from zero every time.

- **Build proposal-to-deliverable pipelines.** Content created at one stage of an engagement can inform others. For example, you might use AI to draft a proposal, then later feed that same text into a report draft to maintain consistency. Or use the project charter text to generate the final executive summary. By setting up a pipeline (even if manually), you ensure work products flow logically and nothing important gets lost. Some firms are even integrating AI into their knowledge management systems so that key insights auto-populate across documents.

- **Use versioning and summarize changes.** When collaborating on deliverables, you can ask AI to compare different drafts and highlight what has changed. This is useful for stakeholder reviews, where you might prompt, *"Summarize how Draft 2 of the report differs from Draft 1."* AI can quickly list, for instance, that Draft 2 *"added a section on competitive analysis; refined recommendations to include timeline; and toned down language on risk X."* This saves time in meetings and emails explaining "what's new" in each version.

- **Create a "visuals library" of reusable AI-generated graphics.** If you've used AI to make a great chart or illustration, keep it for future use. Over time, you'll collect a library of diagrams, icons, and images that you or your colleagues have generated. The next time you need a similar visual, you can pull from the library (or have the AI adapt it) rather than starting over. Just ensure you tag and organize them in a way that's easy to search (e.g., "supply chain diagram—warehouse to customer flow").

Adopting these practices turns AI from a novelty into a dependable coworker that's always by your side. It also helps in scaling your consulting operations. When everyone on the team is using a proven prompt or sharing AI-created templates, you get consistency and collective learning. And importantly, you reduce the mundane grunt work across the board, freeing up more human capacity for client interaction and problem-solving.

AI Saves Time!

How many hours a week do you spend formatting slides or rewriting text you've already explained thoroughly once? How could AI help preserve and reapply that effort instead of reinventing the wheel each time?

In other words, think about where your time is going. If you're doing something repetitive or mechanical, there's a good chance AI can take on that burden. By systematically identifying these opportunities, you can reclaim a surprising amount of time.

Checklist: AI-Enhanced Deliverable Production (a quick recap of best practices):

255

- **Use AI for outlines and section headers.** Kickstart writing by having AI draft an outline or the headings for your deliverable, so you're not facing a blank page or a collection of random notes.

- **Generate one visual per major insight.** Aim to include at least one chart, graph, or illustration for each big idea (many people are visual learners). Use AI to produce or assist with these visuals.

- **Review and refine tone and clarity.** Run all text through an AI writing assistant or editing tool to catch awkward phrasing, jargon, or tone issues. Ensure the deliverable reflects your sensibilities and aligns with your firm's voice.

- **Align content with audience.** Use AI to adapt content for different stakeholders (e.g., create a technical appendix for the engineers, and a high-level summary for the VPs, from the same base content).

- **Archive and tag your drafts.** Store your AI prompts and outputs in a searchable archive (with tags like project type, industry, deliverable section, etc.). This builds institutional memory; you might find that a past AI-generated recommendation for one client provides the impetus for another.

By following this checklist, you create a virtuous cycle. Each deliverable you produce with AI can make the next one even easier, because you'll have accumulated templates, past examples, and refined workflows.

Human-in-the-Loop Quality Control

AI will accelerate your workflow, but it won't replace the consultant's eye for nuance. In fact, the faster pace makes your quality control role even more critical. It's tempting to use an AI-generated report or slide "out of the box" for expedience. But resist that temptation. Always apply a human lens to ensure the deliverable meets the highest standards of accuracy and appropriateness. Judicious use of AI is the name of the game. Treat AI as a junior analyst with superpowers, one that's incredibly fast and wide-ranging, but prone to mistakes and without real-world experience. You, the consultant, are the manager in this relationship. Review everything that goes out the door. In particular, always double-check:

- **Consistency with your voice and brand.** Does the deliverable sound like something you or your firm would say? If the AI uses phrasing that feels off-brand or terminology the client isn't used to, rewrite it. Maintain the consultative tone that builds trust.

- **Accuracy of data and facts.** If the AI generated any numbers, calculations, or factual statements, verify them. AI can and will make errors or even fabrications. Ensure any quantitative analysis is correct (re-run critical calculations yourself) and that any references to research or external facts are valid and current.

- **Validity of insights and logic.** The AI might produce a slick-sounding recommendation that, upon closer examination, doesn't actually make sense. Sanity-check every key message. Ask yourself if it's true, supported by evidence, and whether it reflects appropriate advice. If something strikes you as odd, don't include it blindly. Either fix it or cut it.

- **Relevance of visuals and metaphors.** Auto-generated visuals or analogies also need scrutiny. When AI suggests a clever analogy, consider whether it will resonate with the client or confuse or even offend them. If it creates a chart, does it actually illustrate the point or just look attractive? Ensure every image, graphic, and metaphor in the deliverable serves a purpose and will land well with the intended audience.

- **Clarity for the target stakeholder.** Put yourself in the shoes of your client reader or viewer. Is the language clear for them? If the AI used any undefined acronyms or technical terms, spell them out or simplify. Make sure the narrative flows logically for someone who hasn't been immersed in the analysis like you have.

And a final, crucial point is to protect confidentiality and data security. While using AI, you must be vigilant not to expose sensitive client information. (many firms permit the use of tools like ChatGPT but strictly instruct employees *not* to input confidential client data into them.) Always follow your firm's guidelines and common sense. Anonymize or synthesize data in prompts if needed, use enterprise versions of tools that offer privacy guarantees, and avoid uploading any proprietary materials to consumer AI services.

Your deliverable isn't just a file. It's your proxy in the room when you're not there. Long after you've finished the project, that document or deck may circulate internally throughout your client's organization. (I've had deliverables not only posted for public consumption, but reviewed by the Florida state governor's office and scrutinized by the press.) Make sure it reflects your highest standards. Sweat the details and never let the convenience of AI lure you into complacency. Quality is your responsibility.

In practice, maintaining human-in-the-loop quality means building an AI review step into your workflow. After using AI to generate anything potentially client-facing (even if indirectly), deliberately schedule time to verify and edit that section without AI involvement. Some consultants even print out the AI-assisted report draft and read it on paper to mimic the client experience, catching things they didn't notice on screen. Do whatever it takes, and ensure the final deliverable is something you're proud to put your name on.

Present with Purpose

AI doesn't replace your thinking. It enhances how your thinking shows up on the page or the screen. Done well, AI-assisted deliverables should still sound like you, reflect your logic and structure, and land with even more impact than before. The technology is here to amplify your expertise, not dilute it.

In practical terms, using AI in deliverables can compress timelines for production. What used to take a full day might now take a few focused hours, leaving more time for the relationships, reflection, and strategy that really matter. By automating mundane, routine work, you reclaim capacity to think and to connect with clients.

As you integrate AI into your practice, start small and build confidence. For example, begin by using AI to draft the sections you dread (maybe it's writing a "methodology" section or formatting a list of sources). Use it to polish and proofread text that you've already written (so you don't overthink every sentence). And let it help you visualize ideas that deserve more than bullet points, such as an AI-generated diagram that makes a complex process easily understandable.

At the same time, continue to cultivate the human elements that clients value most, like judgment, empathy, creativity, and trust. An AI might speed up analysis or generate a cool-looking slide, but it won't spontaneously create a breakthrough strategy or build a relationship of trust with your client's CEO. That's your domain.

To summarize this chapter's actionable takeaways:

- **Embrace AI as your creative collaborator.** It can inspire different ways to present information (maybe a narrative story instead of a dry report section) and challenge you to communicate more clearly.

- **Maintain a human-first mindset.** Use AI to free yourself from drudgery so you can spend more time on client interactions, critical thinking, and tailoring insights to client context.

- **Iterate and learn.** Each deliverable you produce with AI offers lessons. Maybe the AI introduced an error you caught, so next time you'll prompt differently. Maybe it saved you an hour on slide design, so next time push that further. Keep adapting your approach.

- **Operate at the highest level of professionalism.** AI is a means to that end, not an excuse to cut corners. Deliverables enhanced by AI should exceed the quality you could have achieved by yourself because you had more time for refinement and review.

The consulting profession has always evolved with new tools, from Excel in the 1980s to PowerPoint in the 1990s to today's generative AI. What doesn't change is the goal of delivering insight and value to clients. AI is just the latest means to help you present with purpose. By embracing it thoughtfully, you're not only working faster, you're opening doors to new ways of crafting impact. In the end, the best consultants will use AI to push the frontier of what great deliverables look like, venturing into new creative territory where others haven't gone yet, and bringing their clients along with them for the exciting journey.

◆

CHAPTER FIFTEEN

Prompting Mastery for Consultants

n the consulting world, knowing how to ask the right questions is at least as important as the answers we provide. And there's no difference when working with AI. When used effectively, AI tools can significantly boost a consultant's productivity and insight. Consultants who integrate AI into their workflow will see the quality of their work improve while raising their task completion rates dramatically, demonstrating tangible efficiency gains. So mastering the art of prompting is more than a tech gimmick; it's a core consulting skill. But these benefits come with challenges. I've encountered consultants who balk at the use of AI, largely due to concerns around trust and accuracy. To harness AI's potential while staying true to a human-first, authentic consulting approach, we need to develop prompting skills that address the consultant's unique needs. This chapter dives into those skills— from understanding the consultant's special prompting challenges, to structuring prompts effectively, refining them iteratively, managing AI memory, and chaining prompts for complex tasks. Throughout, we'll keep a conversational, practical tone and illustrate how consultants can coach AI to better serve their clients.

The Consultant's Unique Prompting Challenges

Every profession using AI will face hurdles, but consultants have a distinct set of prompting challenges shaped by the nature of the work. Unlike a casual user asking a chatbot for trivia, consultants operate in high-stakes, domain-specific scenarios, and the way we prompt must respect that. For example, here are just some of the challenges consultants encounter when working with AI:

- **Context and confidentiality.** Consultants frequently deal with sensitive client information. We can't simply copy and paste raw client data or proprietary reports into a prompt if we're using a public AI service; confidentiality is paramount. This means we often have to abstract or anonymize details when prompting. For example, if you're seeking AI insights on a client's revenue data, instead of using actual figures and names, you should describe the situation in general terms (*"a mid-sized retail client with slowing Q3 sales in the electronics segment"*) to get useful analysis without exposing private data. The challenge is striking a balance between giving the AI enough context to be helpful, yet not violating any confidentiality constraints. As consultants, we must craft our prompts to get substantive answers without giving away sensitive specifics.

- **Need for domain expertise and accuracy.** Consultants are expected to deliver well-researched, accurate insights. However, AI models can sometimes hallucinate (i.e., make stuff up) or produce generic outputs. A consultant's prompt must counteract this tendency by steering the AI toward credible, specific information. For instance, if you're using AI to identify the latest trends in the retail industry, you just might get a disappointingly generic answer—perhaps something about e-commerce growth and the importance of omnichannel—insights any layperson could guess. In that instance, the issue is likely that your query is too broad. Consultants often need tailored analysis (say, trends in luxury apparel retail in Asia post-pandemic), so it's important to refine your approach by adding more context and asking for data-backed points, which will yield far more relevant details. The lesson is that our prompts must reflect the depth of analysis we need. It often helps to explicitly ask the AI to cite data or sources, or to adopt a perspective (e.g., *"from a supply chain expert's point of view"*). We have to remain vigilant; if the AI offers a statistic or fact, double-check it just as you would validate information from a junior analyst. This extra diligence addresses the trust gap, ensuring the AI's contributions can be confidently used in client work.

- **Maintaining authenticity and voice.** Another challenge is making sure the output still sounds like *you* (or your firm) and not some contrived robotic response (which AI happens to be great at providing). Authenticity is a core value in human-first consulting. If every deliverable or email suddenly reads like a cookie-cutter AI template, clients will notice. Thus, a consultant must often prompt in a way that captures the desired tone. For example, you might instruct the AI to use a tone that is *"professional yet conversational, as if explaining to a client"* or *"in the style of our firm's insights, with a touch of humor where appropriate."* In practice, this might mean the difference between an overly formal, generic report and one that feels client-specific and human-crafted. Always includes a prompt phrase like *"write this in a warm, consultative tone"* (or whatever your preferred tone may be) to get a better first draft. Even then, adjust the wording afterwards to ensure it carries your personal voice. The AI is a tool to amplify our thinking, but not a substitute for our personal touch. The prompting challenge is to get helpful content out of the AI while still leaving room to infuse our own insights and style.

- **Evolving workflows and habits.** Incorporating AI into consulting work requires a shift in how we approach tasks, which involves changing old habits (and we all know how easy that is). A seasoned consultant may be used to scouring databases or brainstorming on a whiteboard. Now, they might turn to an AI assistant for a quick literature scan or a first pass at brainstorming ideas. Adapting to this new workflow can feel awkward at first. It takes discipline and practice to learn how to "talk" to the AI effectively and to integrate it smoothly into our work routine. For example, you might initially find it time-consuming to think of the perfect prompt. Over time, as you gain prompting mastery, you start to develop an intuition for what details to include and what style of question gets the best results. The mindset to adopt is that of a coach or mentor to the AI, patiently guiding it with clear instructions, and iteratively improving its output. This mirrors how we might coach a junior consultant; we wouldn't expect them to produce a flawless analysis on the first try without guidance, and the same goes for an AI assistant.

Despite these challenges, consultants are quickly learning that effective prompting is worth the effort. With just a little training, even basic chatbot usage can yield massive performance boosts in consulting tasks. Let AI handle those tasks we're not great at doing so we can excel at the ones we are. In other words, use AI to take care of the grunt work or first drafts, while you focus on the high-value judgment calls and creative solutions. The rest of this chapter will explore how to do that in practice, starting with how to structure an effective prompt.

Structuring Effective Prompts

A well-structured prompt is the foundation of getting useful output from AI; the more specific and well-organized your guidance, the better the result. As OpenAI's own prompt engineering guide emphasizes, being specific, descriptive, and detailed about what you want—the context, the outcome, the format, the style—significantly improves the quality of responses. Rather than typing the first question that comes to mind, it pays to take a minute and structure your prompt. Here's how consultants can craft effective prompts, step by step:

- **Start with context or role.** Begin by setting the scene for the AI. If the model knows the background or the role it should take, it can provide a more relevant answer. For example, you might start with, *"You are an expert management consultant advising a client in the automotive industry facing declining sales."* This gives the AI a persona and context to work with. Similarly, if you're giving it text or data to work on (e.g., asking for a summary of provided text), put any instructions at the very beginning and clearly separate them from the content. One recommended approach is to use delimiters like triple quotes to isolate the context. For instance, *"Summarize the findings below in bullet points: [next line] """ [Client report text goes here] """."* Placing the instruction up front and using quotes or separators prevents confusion between what's instruction and what's reference text. In short, lead with the "what" and the "who"—what you want it to do, and whose perspective it should adopt.

- **Be clear and specific in your ask.** Vague prompts lead to vague answers. Frame a clear question or task for the AI. If you need an analysis, be specific. For example, instead of asking *"What about the competition?,"* ask *"What are three key competitive advantages of [Client X] over [Competitor Y] in the European market?"* Include details that focus the AI, such as the number of points you want, particular aspects to emphasize, time frames, etc. Provide some context, then a concise question to direct the AI, and use follow-up prompts to drill down further. The initial prompt should effectively define the problem for the AI in unambiguous terms. Imagine you're writing the question for a seasoned expert, where you give sufficient detail to communicate exactly what you're after.

- **Specify the desired output format and style.** If you want the answer in a list, a table, a narrative, or a slide outline, say so explicitly. You can even give an example of the format. For instance, *"Provide the answer as a concise bulleted list of five or six items."* If tone or style matters, include something like, *"Use a formal tone appropriate for a board presentation,"* or *"Explain it as you would to a non-technical client."* Consultants often need polished outputs that can be client-ready, so guiding the style is important. Consider constructing AI instructions with phrases like *"Provide a concise, bullet-point summary,"* or *"Act as a finance expert and use credible sources in your answer."* By front loading such instructions, you get responses that are much closer to the mark. The AI will mimic the format you ask for whether it's an email draft, a SWOT analysis, or a friendly Q&A. Don't be afraid to practically script the structure (e.g., *"First, give an introduction, then present three recommendations numbered 1, 2, 3, with a brief rationale for each."*). It may feel pedantic, but these models perform far better when provided with detailed, unambiguous instructions.

- **Include examples if needed.** If your task is complex or you have a very particular style in mind, giving a brief example can help. This is known as providing a one-shot (one example) or few-shot (several examples) prompt. For example, if you want the AI to produce an interview-style dialogue, you might show a short example of the Q&A format before asking it to generate new content. Examples act as templates. However, for many consulting uses, you might not need to go that far; often a well-worded instruction suffices. Start simple, and only if the AI isn't getting it, consider adding an example.

- **Double-check and refine the prompt.** Before hitting enter, quickly review your prompt. Ask yourself if it clearly tells the AI what you need, if any crucial detail is missing, or whether it could be misinterpreted. For instance, the prompt *"Draft a brief on market entry"* is open to interpretation: brief of what, for whom? A quick addition like *"...brief for a tech startup entering the EU market, covering regulatory hurdles and competitive landscape"* makes it far clearer. This small adjustment to your prompt can save you from having to ask again. Think of it as debugging your query. It's often helpful to imagine how an overly literal minded person might misread your request, then preempt that by clarifying the wording.

To illustrate these principles, consider a before-and-after example. A "before" prompt such as *"Tell me about the retail industry outlook"* will likely yield a rambling answer full of generic facts. An "after" prompt, like *"You are a consulting analyst researching the retail industry outlook for 2026. Identify three major trends shaping retail (e.g., consumer behavior, technology, or supply chain), and for each trend provide one concrete example or statistic illustrating its impact. Present the answer as three bullet points with a brief explanation for each, in a neutral analytical tone."* The difference in the AI's response will be stark. The revised prompt will produce a focused list of trends (like *"E-commerce penetration continues to rise, e.g., online sales now ~25% of all retail sales in 2025"*, etc.), each with a relevant data point. The effort spent structuring the prompt will pay off in efficiency; less back-and-forth will be needed to get a useful answer.

Structuring effective prompts might feel formulaic at first, but soon it will become second nature. By consistently providing context, clarity, and guidance on format, you essentially coach the AI to deliver what you want. The key is to remember that the AI cannot read your mind—it only reads your prompt, so give it the right cues. Next, we'll look at what to do when the output isn't quite right on the first try, and how to iteratively refine prompts to zero in on the perfect answer.

Prompting as a Dialogue

One of the most powerful aspects of modern AI tools (like ChatGPT and its peers) is that interaction is conversational and iterative. You're not limited to asking one single perfect question; you can engage in a back-and-forth, refining the AI's output with each round. Think of it as a dialogue where you gradually steer the AI toward a better answer. Prompting mastery involves embracing this iterative process rather than expecting a perfect response on the first try.

In practice, iterative prompt refinement involves reviewing the AI's first answer critically, then tweaking your prompt or asking follow-up questions to improve the result. It's very much like how we work with junior team members; rarely is the first draft of an analysis or slide deck final. We give feedback, clarify requirements, and ask for revisions. Similarly, with an AI, you might start with a decent but not great answer and then say, *"Okay, now can you make it more concise?"* or *"Explain that in simpler terms,"* or *"Can you provide an example to back that point?"* Each prompt in the sequence hones the response further.

Research and practical guides on prompt engineering echo the value of this approach. By iteratively adjusting prompts by adding constraints, clarifying terms, or including new information based on the previous output, you align the results more closely with your goals, reduce errors, and achieve more consistent quality. In fact, an iterative process encourages a form of feedback loop. The AI's output is feedback on your prompt, which you then use to refine your next prompt. This loop continues until you're satisfied. Start with a solid prompt, review the output for sufficiency, modify and repeat until the response sounds right. Following this cycle will yield better outputs aligned with your objectives and help you to catch issues before you've gone too far down the wrong path.

Let's walk through a realistic consulting scenario to illustrate iterative prompting. Suppose you're preparing a proposal for a client looking to improve their customer service operations, and decide to use AI to brainstorm some initial ideas and best practices. You might start with a prompt like, *"Give me some ideas to improve customer service for a telecom company."* The AI might respond with a few generic suggestions (e.g., *"train employees," "use a CRM system," "gather customer feedback"*). It's a start, but too high-level. In reading the output you realize you need more specificity. So you engineer your follow up prompt to state, *"These are a good start. Now, for each idea, can you add a concrete example of how a company might implement it?"* The AI will oblige and add details (like an example of a telecom firm implementing a new AI chatbot to handle support calls, under a "use a CRM/technology" idea). Now the suggestions have more heft.

You might then notice the AI still hasn't touched on metrics or process streamlining, which you deem to be important topics, so you refine your prompt further by stating, *"Also include ideas related to process improvement or metrics tracking in customer service, if not already mentioned."* This prompt nudges the AI to broaden its answer, so the next output might include a point about *"streamlining call scripts and measuring first-call resolution rate,"* which is exactly the kind of operational detail you might have in mind. With a few rounds of prompting, you'll get a robust list of tailored suggestions.

A few tips become evident from this example. First, don't hesitate to ask follow-ups. The AI doesn't get annoyed or tired; it's there to help until you're satisfied. Second, you can reference the AI's previous answer in your follow-up prompt. Phrases like *"elaborate on point two"* or *"rephrase the last paragraph to be more client-friendly"* directly build on the conversation. The model remembers what it just told you (within the limits of its memory, as we'll discuss in the next section), so you can iteratively zero in on what you need. Third, know when to stop. Iteration is great, but there's a point of diminishing returns. If you find that each refinement is only making tiny improvements, you probably have enough to work with. At that stage, it might be more efficient to take the AI's output and polish it yourself with your expert judgment and tone.

One more advanced iterative technique is the self-critique approach, where you ask the AI to critique or improve its own answer. For instance, *"Review the above answer and check for any gaps or incorrect assumptions. Update the answer if needed."* Surprisingly, the AI will often catch something it missed and refine the answer. This works because you're essentially prompting it to reflect and iterate on its own output.

Remember, iterative refinement is a collaborative process between you and the AI. It exemplifies the coaching mindset—you're guiding the AI like a coach or editor would, step by step. And just as in human collaboration, communicating clearly and giving constructive feedback (via your prompts) leads to better outcomes. Prompting mastery isn't about getting it perfect in one go; it's about knowing how to evolve a prompt based on results, which is exactly what makes AI a powerful partner for consultants. In the next section, we'll tackle a practical constraint that can arise in longer AI interactions—the model's memory and context window.

Memory and Context Windows

As you engage in multi-step dialogues and feed background information to the AI, you'll eventually bump into the limits of its memory—technically known as the model's context window. The context window is essentially how much text the AI model can handle at once. Think of it as the AI's working memory or the size of its "mental workspace." Today's AI models have impressive but not infinite memory. For example, early GPT-3 models could handle about 4,000 tokens of text (roughly 3,000 words) in their context. Newer models have expanded that to hundreds of thousands (more than 1 million in some instances) of tokens. But even 100,000 tokens (about 75,000 words) is still somewhat limiting—considerably less than the text of this one book. In consulting terms, you might fit a large proposal or several documents into your prompt, but your entire knowledge library will take quite a bit more.

The key takeaway is that AI cannot remember an unlimited amount of information from earlier in the conversation. Once you exceed the context window, older parts of the conversation (or long documents you fed in) begin to drop off; the model literally no longer "sees" them in its input. Even before hitting the absolute limit, there is something called the "murky middle" problem—when the context gets very large (tens of thousands of tokens), the model might pay less attention to details in the middle of that context. It tends to grasp the gist of the entire input but could gloss over a specific detail buried on page twenty of a fifty-page document. In practical terms, if you dump a huge amount of data into a single prompt and ask a very narrow question about it, the AI might miss the needle in the haystack. It's not that it deliberately ignores it, but these models are optimized to handle general patterns and may not perfectly memorize every fine-grained fact in a long input.

So, how do we work within these memory limits? Readers will likely recognize some common consulting that map well to AI use:

- **Chunking and summarizing.** If you have a large body of text or data, don't feed it all at once. Break it into smaller chunks and interact on those. For instance, if you have a 120-page report to get insights from, you might prompt the AI with the first ten pages asking for key themes, then do the next ten pages, and so on. Afterward, you can take those summaries and ask the AI to synthesize the highlights across the summaries. This approach not only keeps each interaction within the context window, but it also allows you to verify the output in stages. In fact, some consultants suggest feeding data in increments specifically to ensure you can check the AI's output at each step and maintain quality control. This incremental approach is essentially a manual form of *retrieval augmentation*; you digest info in pieces rather than one gulp, which reduces the chance of the AI misunderstanding or forgetting critical bits.

- **Refresh or reset the context.** In a long brainstorming session with an AI, you might notice it starts to lose track of earlier details or the style drifts. One tactic here is to periodically summarize the conversation so far and use that to start a new session or as a new prompt. For example, *"To recap, we have discussed A, B, and C. Now let's move on to D."* By providing a succinct summary in a new prompt, you effectively compress the important context into a smaller package that fits in the window. Many AI chat interfaces do some version of this automatically (for instance, when you open a new chat, you might copy the last relevant answer as context). You can do it manually to be safe. Similarly, if you realize you've gone on a tangent, it might be useful to start fresh with the relevant bits rather than dragging the entire history along.

- **Use distinct sessions for distinct tasks.** An insight from practitioners is the concept of keeping separate workspaces for different topics. If you're working on two unrelated analyses (say, one prompt thread for a client's market entry strategy, another for brainstorming an internal team meeting agenda), it's wise to use separate chats or sessions or, as some AIs permit, separate "projects" complete with instructions and sample data or documents, for each. This avoids the AI mixing contexts (no risk of it pulling in the meeting agenda info while discussing market strategy). Context management is critical. Be sure to keep different AI chat sessions for different focus areas to keep the conversations clean. By doing this, you ensure each context window is dedicated to one stream of thought. It's like having different folders for each project; it keeps things organized and reduces the chance of confusion.

- **Explicitly reminding the AI of key facts.** If certain details are crucial and you suspect the model might "forget" them as the conversation grows, don't hesitate to remind or re-provide those details in later prompts. For instance, *"Recall that the client's budget is only $100,000, so any solution must be low-cost."* That way, even if that fact was stated much earlier, you bring it back into the active context. Yes, this can feel redundant, but until models have truly expansive or long-term memory (and they're getting there!), redundancy can be a feature, not a bug. It ensures important constraints stay in view.

Understanding context windows is also important when deciding what not to stuff into a prompt. If you load your question with every single piece of information (hoping to cover all bases), you might be wasting precious context space on details that don't actually change the answer. It's a bit like having a meeting and inviting twenty people when only five really need to be there, causing the conversation to become unfocused. A skillful prompter includes the necessary context (enough for the AI to be informed) but not so much that the core question is drowned out. For example, if you want a summary of a report's implications, you might not need to include the entire methodology section of that report in your prompt. The executive summary or specific data points you want analyzed will do.

In summary, memory and context are finite resources in an AI consultation. We, as consultants, have to manage those resources wisely, much like we manage our own time or our team's workload. By chunking inputs, resetting context when needed, separating concerns, and highlighting key info, we can work smoothly within the AI's memory limits. And if you ever hit a wall (e.g., the AI seems to be forgetting earlier discussion topics), you now have strategies to get past it. With memory management in hand, we can turn to a closely related advanced skill, prompt chaining, which involves structuring multi-step interactions to tackle complex tasks.

Prompt Chaining for Complex Tasks

Some consulting questions are straightforward enough to answer in one go, but many are multi-layered or complex. Think about something like, *"Develop a market entry strategy for Client X in Country Y."* That's not a single question; it's a bundle of tasks (market analysis, competitor assessment, regulatory considerations, entry tactics, etc.) all rolled into one. If you naively throw that entire request at an AI in one prompt, you'll likely get a superficial answer that touches a bit of everything but without real depth. This is where prompt chaining comes in. Prompt chaining is the technique of breaking down a complex task into a series of smaller, linked prompts, where each prompt addresses a subtask and builds toward the final result.

In essence, prompt chaining allows you to guide the AI through a problem step by step, rather than expecting it to deliver a grand solution in one shot. This approach has multiple benefits. For one, it often improves performance because the AI can focus on one aspect at a time with clarity. It's the difference between asking someone to write a business plan from scratch in five minutes vs. asking them first to outline key sections of a business plan, then fill in details section by section. By chaining, you also gain transparency and control; you see the output of each stage and can adjust if something seems off. It's much easier to pinpoint where a mistake or irrelevant tangent occurs if each prompt handles a defined part of the task. Without chaining, if the middle part of a response is wrong, you'd have to debug the entire thing. With chaining, you can fix the specific sub-prompt that caused the issue. This is analogous to modularizing a project, i.e., breaking a big project into manageable pieces makes it simpler to manage at each step.

Let's go through a consultant-oriented example. Let's say you need to use AI to help formulate that market entry strategy for Client X in Country Y. You decide to chain prompts as follows: In the first prompt, you ask, *"What factors should be considered when entering the [Country Y] market for [Client X's industry]?"* This is a broad question aimed at gathering the key dimensions (market size, competition, regulations, consumer preferences, etc.). The AI returns a list of factors, which you review and find to be pretty solid—perhaps it lists things like local customer behavior, major competitors, legal barriers, distribution channels, and so on. Now, in the second prompt you take one of those factors (e.g., competitors) and dive deeper with, *"Who are the top five competitors in [Client X's industry] in [Country Y], and what is a brief overview of each (market share, strengths)?"* The AI provides a rundown of competitors. This is valuable intel you'll use to develop the strategy. In the third prompt, you might focus on another factor, like regulations, with, *"Summarize any important regulations or government policies in [Country Y] that could impact a new entrant in this industry."* Again, the AI fetches relevant points (perhaps about licensing requirements, tariffs, etc.). With the fourth prompt, you start synthesizing. With the info gathered, you might ask, *"Given the above factors (market environment, key competitors, regulations), suggest two or three market entry strategies for [Client X] in [Country Y], and for each, briefly state why it would be effective or risky."* Because the AI "remembers" the prior conversation in the chat, it can use the specifics it listed (like competitor strengths or regulatory points) in crafting these strategy suggestions. The output might be something like, *"Strategy 1: Partner with a local firm (because competitor analysis showed local players have distribution networks... etc.); Strategy 2: Niche*

market focus (leveraging a gap identified in consumer preference data)..."
and so on. Finally, with a fifth prompt you could even ask the AI to turn
those points into a mini-slide outline or executive summary paragraph that
you can directly integrate into a proposal.

Prompt chaining allows you to divide and conquer the problem. Each chain
link (prompt) had a clear purpose and makes the final output more robust. If
at any stage you get an unsatisfactory answer, you could refine that *before* it
polluted the final strategy. For example, if the competitor list missed a
known player, you could correct that in the second prompt's result by explicitly asking, *"Are there any important regional competitors besides those
listed?"* and the AI could update the list. This beats getting all the way to the
final strategy and then realizing a key competitor was overlooked, forcing
unwanted and inefficient rework.

Prompt chaining is not only useful for strategy cases. It's a general approach
that works whenever you have a complex task and ask yourself how it could
be logically broken into parts. Another common use case in consulting is
document analysis or Q&A. Suppose you have a huge document and a tough
question to answer about it. A chaining approach might involve prompting
the AI first to extract or highlight relevant sections of the document that per-
tain to the question (essentially, *"find the info that might help answer this"*).
Then, in a second prompt, ask it to formulate an answer based on those ex-
tracted pieces. This two-step chain (find relevant info, then use relevant info
to answer) often yields a more accurate and grounded answer than a one-
step "answer in one go" prompt, because the model focuses on retrieval first,
reasoning second. In fact, that's a mini-version of how advanced AI systems
do RAG (Retrieval-Augmented Generation); they first gather facts, then
generate answers.

Chaining can also help enforce methods or frameworks. For example, you
could prompt an AI to perform a SWOT analysis in steps with, *"List
Strengths of X"*, then *"List Weaknesses"*, and so forth, then finally *"Combine into a SWOT summary."* This ensures each category is well thought out.

When using prompt chaining, keep these tips in mind:

- **Maintain coherence between prompts.** Each subsequent prompt might rely on information from earlier ones. Make sure the AI has that context. In a chat setting, the history usually provides it. If you're using separate prompts outside of a single conversation (say, in a script or separate API calls), you'll need to feed the relevant outputs back in.

- **Be mindful of context window.** The more you chain in one conversation, the more you might approach the context limit, especially if each answer is long. If you notice the AI starting to forget earlier parts in a long chain, it might be time to summarize or trim, as discussed in the memory section. Sometimes it can help to consolidate interim results before moving on (e.g., *"Summarize the competitor analysis in one paragraph"* before asking the final question, to reduce the text it needs to hold in memory).

- **Flexibility to pivot.** The beauty of chaining is you can pivot if a path isn't fruitful. If by the second prompt you realize the approach needs work, you can change course. You're not locked into a single massive prompt. Treat it as an interactive exploration.

By chaining prompts, you essentially create a structured dialogue that mirrors a consulting thought process. It encourages you, the consultant, to think in a structured way as well by considering what you need first, what next, and so on. This often leads not only to better AI output but also to clearer thinking about the problem itself. It's like the AI is helping to facilitate a structured problem-solving session with you.

To conclude this section, prompting mastery for complex tasks is about strategy. You strategize your prompting in the same way you strategize your consulting approach. You wouldn't attack a client problem without breaking it into manageable workstreams; likewise, don't hesitate to break a prompt into manageable sub-prompts. The end result is a chain of AI contributions that, combined with your own analysis, can produce a truly high-quality deliverable.

In mastering prompting, we've covered understanding the consultant's unique challenges, crafting effective prompts, refining through iteration, managing context memory, and chaining prompts for complexity. Throughout all of this, the golden thread is maintaining a human-first, authentic consulting mindset. AI is a powerful amplifier of your skills, but *you* remain the director of the engagement. Use prompts to guide the AI like you would guide a talented intern: provide clear instructions, iterative feedback, and a structured approach to obtain the best outcome. And always filter the AI's work through your human lens of expertise, ethics, and empathy. When done right, AI prompting becomes a natural extension of your consulting toolkit, one that lets you focus more on the creative, strategic parts of your work that truly add value, while your AI assistant helps cover the balance. By practicing these prompting techniques, you're actually *collaborating* with AI. In the evolving landscape of consulting, that collaboration will be a key differentiator, enabling you to deliver insights and results faster and perhaps better than ever before, all while preserving the authentic, human touch that clients trust.

Afterword

When I began writing this book, the pace of change in artificial intelligence already felt dizzying. By the time you are reading these words, it's entirely possible that some of the tools, techniques, or even the terminology I've described have been surpassed. That's the nature of a technological revolution—it doesn't wait for anyone to catch up, and it certainly doesn't pause long enough for books to remain perfectly current.

This reality might tempt some to dismiss efforts like this as futile. Why codify ideas that may be out of date in months, or even weeks? My answer is simple: while the tools will change, the principles will not.

What matters most is not whether you are using ChatGPT, Claude, Gemini, or whatever AI tool or model emerges tomorrow. What matters is how thoughtfully you integrate these capabilities into your professional practice. Consultants, advisors, managers, and leaders thrive not by memorizing the latest toolset but by understanding how to frame questions, structure engagements, and deliver value to clients. AI will accelerate, amplify, and in some cases redefine those activities, but it will not remove the need for judgment, empathy, and courage.

The professional services world is built on trust. Clients come to us because they believe we can help them solve their most pressing problems. The worst mistake we could make in adopting AI is to use it as a shortcut that erodes that trust, outsourcing not only the mechanics of our work but also the thinking, the care, and the accountability. AI can draft, summarize, simulate, and suggest. But only you can decide what's worth saying, which risks are acceptable, and what "good enough" really means for a client.

At the same time, there is enormous opportunity for those who lean in constructively. The consultant or manager who learns to harness AI responsibly will be able to operate at a level of speed and breadth unimaginable just a few years ago. Analyses that once consumed weeks can be completed in days. Drafts that once felt like laborious starting points can now emerge in minutes. Brainstorming that might have stalled in a small room can be supercharged by an AI partner that never tires of offering new angles. And that book you've always wanted to write? AI can finally get you there. This isn't about replacing human expertise; it's about expanding what human expertise can achieve.

Yes, some of what you've read here will inevitably become outdated. A new feature will arrive tomorrow that makes today's process obsolete. A breakthrough in reasoning, multi-modal understanding, or agentic coordination will shift the field yet again. But the broader lesson remains steady: don't chase shiny objects. Anchor your use of AI in your authentic consulting style, in your ethical commitments, and in the outcomes that matter for your clients. If you do that, then even when the particulars shift, your foundation will remain strong.

As I close this book, I want to highlight one more truth: AI adoption is not just about efficiency. It's also about creativity. When you let go of the mechanical tasks that once consumed you, you create space for deeper thinking. You can devote more energy to strategy, to relationship-building, to vision. That's where the future of professional services lies—not in competing with machines, but in designing human-centered experiences that machines help to support.

We are at the beginning of a long journey. The last few years have been an opening act. The next decade will bring transformation at a scale we can barely imagine. Those who resist, hoping the storm will pass, may find themselves stranded. Those who embrace AI uncritically, hoping it will solve everything, may find themselves disillusioned. But those who approach it with curiosity, discernment, and courage will discover an extraordinary ally.

That is my hope for you: that you see this book not as a manual frozen in time, but as a companion in your own evolving practice. Use it to ground your approach. Question it when the details no longer fit. Adapt it as new capabilities emerge. And most of all, carry forward its central conviction— that AI, thoughtfully applied, can make us not less human but more effective, more creative, and more valuable in the work we do.

The age of artificial intelligence is here. The question now is not whether it will shape our profession, but how we will shape our profession with it. That responsibility—and that opportunity—belongs to you.

ROB BERG

◆
_____

Prompt Library by Role and Use Case

I
n the modern consulting landscape, artificial intelligence has become a trusted ally at every step of the consulting process. This chapter provides a prompt library organized by consulting role and use case, illustrating how AI-driven prompts can augment your workflow. We will explore key categories, including Market Research, Proposal Development, Client Onboarding, Risk Analysis, Thought Leadership, Workshop Design, and Executive Briefing, each framed with narrative context that highlights why these prompts matter. Throughout, the emphasis remains on a thoughtful, human-centered approach: using AI creatively and curiously to empower both you and your clients. The goal is not to replace the consultant's judgment, but to enhance it, allowing you to differentiate yourself through authentic insights and personalized client service.

As you read through each category, imagine how these prompts function as a toolkit for a solo consultant or a small firm. They enable you to punch above your weight, executing tasks with the efficiency of a larger team while maintaining the personal touch that defines your brand. We'll also discuss strategies for customizing prompts to client-specific needs and training your clients to confidently use AI-enhanced tools themselves. By the end of this appendix, you should feel not only informed about various prompt use cases, but also inspired to experiment with them—guided by creativity, curiosity, and a clear sense of purpose.

Market Research

Market research is a foundational step in many consulting engagements. It's the phase where curiosity reigns: you're exploring a new industry, an emerging trend, or a competitive landscape to gather insights that will shape your recommendations. In the past, thorough market research required laborious data gathering—combing through reports, financial statements, news articles, and databases. Today, AI can function as your tireless research assistant, ready to scan vast information sources and summarize key findings in seconds. For a solo consultant, this means being able to ramp up quickly in a domain, almost as if you had an entire analyst team on call. The following prompts in our library will help you harness AI to gather industry intelligence, identify trends, and understand competitive dynamics.

Using AI for market research allows you to ask broad questions and then progressively drill down. You might begin with an open-ended exploration of an industry's outlook, then pivot to specific questions about consumer behavior or competitor strategies. Well-crafted prompts can direct the AI to pull relevant insights from multiple sources—for example, analyzing news articles, blogs, and social media to detect emerging themes. This capability means you can uncover signals that might otherwise remain hidden under an avalanche of information. Want to know the latest regulatory changes affecting your client's market? Or the sentiment of customers toward a new product category? With the right prompt, an AI tool can sift through the noise and present you with a coherent summary of findings. It's an empowering feeling to have such breadth of information at your fingertips, fueling your creativity as you formulate hypotheses and strategies.

However, a human-centered consultant remains mindful of AI's limitations. While generative AI excels at connecting dots across data points, it isn't infallible. The knowledge it provides is largely based on existing data, which means it may not have up-to-the-minute information and can sometimes misinterpret correlation as causation. For instance, the AI might detect a pattern in online discussions and suggest a trend that looks significant, but it's your job to verify whether that trend is real or just a quirk of the data. Bias is another concern: if the underlying sources are skewed, the AI's output will be too. As you use prompts from this library, remember to double-check AI-generated insights against trusted sources and your own expert judgment. The true power of AI in market research emerges when its speedy analysis is combined with your critical thinking. Use the AI to cast a wide net and surface possibilities, then apply your consultant's intuition to validate and refine those insights. In this way, you ensure that technology serves your pursuit of truth and relevance, rather than leading you or your client astray.

Example Prompts for Market Research:

- *"Provide an overview of the current trends and growth projections in the [specific industry] market. What key factors are driving change in this space?"*

- *"Analyze the top 5 competitors in [client's target market]. Summarize each competitor's market share, strengths and weaknesses, and recent strategic moves."*

- *"Scan recent customer reviews and social media posts about [product/service]. What common pain points or desires are customers expressing, and how might these indicate gaps in the market?"*

Using prompts like these, you can quickly gather a broad understanding of the market landscape. The AI will help highlight important patterns and data points, which you can then investigate further or use as a basis for deeper analysis.

Having gathered critical market insights, you'll be well-prepared to tackle the next step: translating that knowledge into a compelling proposal. The transition from research to proposal writing is where information meets persuasion. How do you weave facts into a narrative that wins the client's confidence? In the next section, we'll explore how AI can assist in Proposal Development, ensuring your recommendations are not only well-informed but also communicated with clarity and impact.

Proposal Development

Crafting a winning proposal is both an art and a science. On one hand, it demands clarity, structure, and attention to detail—laying out the project scope, approach, timeline, and benefits in a way that resonates with the client's needs. On the other hand, the best proposals tell a story, conveying a narrative of how your solution will uniquely solve the client's problem. AI can support you on both fronts. With the prompts in this section of the library, you can use AI to generate structured outlines, suggest persuasive language, and even infuse creativity into your proposals. For a small consulting firm or solo practitioner, this means no longer staring at a blank page wondering how to start. Instead, you can collaborate with your AI assistant to produce a solid draft in a fraction of the time, which you can then refine with your personal expertise and style.

When using AI for proposal development, begin by outlining the core elements: What objective are you addressing? What is your proposed solution and methodology? Who will do the work, and in what timeframe? You can prompt the AI to *"Draft a proposal outline for [project description], including sections for objectives, methodology, deliverables, timeline, and ROI."* The result is often a clear scaffold that ensures you haven't missed any key pieces. From there, you might ask the AI to elaborate on certain sections. For example, *"Give me a persuasive paragraph on the business benefits of adopting [specific solution], aimed at a client in [industry]."* The AI can help articulate value propositions and even suggest analogies or examples to strengthen your case. Proposal writing requires clear communication and strategic showcasing of benefits—areas where AI-generated text can be a helpful starting point.

However, while AI can produce competent prose, it's the human touch that wins deals. As a consultant, you bring insights that no generic model can replicate, such as knowledge of the client's unique context, your personal credibility, and an authentic voice. It's crucial to infuse those human elements into the final proposal. In fact, if you rely solely on AI's output, you risk your proposal sounding generic or interchangeable with others; always add your own spin, your human touch—differentiation through authenticity is key. Use the AI to handle the heavy lifting of creating a first draft or suggesting phrasing, but always review and tailor the content. Make sure the proposal speaks directly to the client's situation. Reference their goals, use their terminology, and include anecdotes or case studies from your own experience where appropriate. The prompts in this library might, for instance, help you generate multiple ways to explain a concept; you can then choose the one that feels most genuine and effective. Ultimately, proposal development is where your expertise shines through. AI accelerates the process and sparks ideas, but your personal touch and strategic thinking convert a good proposal into a winning one.

Example Prompts for Proposal Development:

- *"Outline a consulting project proposal for [client scenario], including problem statement, proposed solution, project phases, deliverables, and success metrics."*

- *"Help me write an engaging executive summary for a proposal to [client type] that emphasizes how our approach will deliver [specific benefits]."*

- *"Suggest three different ways to phrase the value proposition of our solution (for example, focusing on cost savings, growth opportunities, or risk reduction) so I can choose the best wording."*

These prompts can kick-start your proposal writing process. The AI's responses will give you structured content and persuasive language to work with. From there, you'll refine each section, ensuring the tone and details are perfectly aligned with your client's expectations.

With a polished proposal in hand, you secure the engagement—congratulations! But winning the work is only the beginning. Next comes the crucial phase of client onboarding, where you set the stage for a successful project. Just as AI helped in research and proposal writing, it can also assist in creating a smooth and professional onboarding experience. Let's explore how.

Client Onboarding

Once the contract is signed, the focus shifts to onboarding the client and laying the groundwork for project success. This phase is all about building trust, clarifying expectations, and getting organized. As a consultant, you want to make a strong first impression in these early interactions, demonstrating professionalism and empathy. AI-enhanced prompts can support you by generating checklists, kickoff agendas, communication plans, and other

onboarding materials quickly and thoughtfully. For smaller consulting practices, this means you can deliver a "big firm" onboarding experience without
a large support staff, ensuring no detail falls through the cracks.

Client onboarding typically involves gathering essential information from
the client, aligning on project objectives, and establishing communication
norms. For instance, you might use prompts to help draft a welcome email
or a kickoff meeting agenda. A prompt like *"Create a kickoff meeting agenda
for a new project with a client in [industry], including introductions, project
overview, roles and responsibilities, timeline review, and next steps"* will
yield a structured plan that you can tweak to fit your specific situation. You
could also ask the AI to *"List the key pieces of information to request from
the client during onboarding (e.g., relevant data, access to systems, key contacts)"* to ensure you have a comprehensive intake checklist. By covering
these bases, you show the client that you are thorough and proactive. Consultants who communicate clearly and anticipate client needs from the start
are more likely to earn client confidence. In fact, consistent and polished
communication is known to build trust with clients, fostering strong client-
consultant relationships. Using AI to draft well-worded, organized onboarding documents helps you maintain that polished communication, even when
you're moving fast.

Another valuable use of AI during onboarding is customizing materials to
the client's context. Perhaps you have standard templates for project charters
or stakeholder questionnaires. You can feed these into the AI with a prompt
to tailor them, such as, *"Here is my project charter template. Help me customize the tone and content for a client who values [e.g., a collaborative
approach or a very data-driven approach]."* The AI can adjust wording to
align with the client's culture (for example, more formal vs. more casual, or
highly detailed vs. high-level summary). This ensures that from the very first
deliverables, the client feels "this was made for us," reinforcing that human-
centered touch. Remember, onboarding is not just about paperwork, it's
about setting a tone for collaboration. So while AI can generate the drafts,
it's wise to review them with an eye for personal touches like mentioning
the client's mission or referencing a comment someone made in earlier meetings, which can make the communication feel warm and bespoke.

Efficient onboarding aided by AI also frees you to focus on the human side of the relationship. Instead of spending hours manually compiling orientation documents, you can invest that time in conversing with the client's team, understanding their unspoken concerns, and fostering a sense of partnership. Empathy and personal connection, amplified by the organization and clarity that AI-assisted prep work provides, create a powerful combination. By the time onboarding is complete, the client should feel confident that they've chosen the right consultant—one who is both technologically adept and personally attuned to their needs.

Example Prompts for Client Onboarding:

- *"Generate a list of information and access points we should request from the client during onboarding (e.g., relevant documents, key contacts, software access, etc.), tailored for a project in [client's industry]."*

- *"Draft a friendly yet professional welcome email to the client's team, outlining the next steps now that our consulting engagement is kicking off."*

- *"Create a communication plan template for this project, specifying how often we will update the client, in what format (email, calls, meetings), and what each update will include."*

Prompts like these help ensure you cover all onboarding essentials. The AI can quickly produce well-structured lists and messages, which you can then personalize. By presenting the client with organized plans and clear communication right from the start, you establish credibility and trust.

With the project launched on solid footing, the next step is to navigate the inevitable uncertainties that come with any complex endeavor. Consulting projects often face unknowns and potential pitfalls—which is why Risk Analysis is so important. Proactively identifying and addressing risks can set your project apart and demonstrate foresight. In the following section, we'll see how AI-driven prompts can assist you in anticipating challenges and planning mitigations, giving both you and your client greater confidence as you move forward.

Risk Analysis

In consulting, being prepared for the unexpected is part of delivering value. Risk analysis is where you channel your curiosity into probing "What could go wrong?" and your creativity into devising safeguards. Traditionally, risk management might involve brainstorming sessions, reviewing past project post-mortems, or consulting risk checklists. These remain valuable techniques, but AI can turbocharge the process by systematically generating potential risk scenarios and mitigation ideas that you might not immediately think of. Especially for a solo consultant without a team of specialists, an AI prompt can serve as a brainstorming partner, ensuring you consider a broad spectrum of risks, from market fluctuations to operational hiccups to stakeholder issues. The prompts in this section of the library will help you uncover hidden landmines in a project plan and develop thoughtful responses to them.

To use AI effectively for risk analysis, start by asking it to enumerate risks in a given context. For example: *"List the top 10 potential risks for a project implementing [specific solution] in [industry or environment]."* The AI will draw on its knowledge base to produce a list that could include technical risks, financial risks, compliance risks, and more. You'll likely see some familiar items on the list, along with a few novel ones that merit consideration. From there, you can take each identified risk and delve deeper: *"For the risk of [risk X], suggest possible mitigation strategies and contingency plans."* This approach leverages the AI's ability to retrieve lessons from countless case studies and scenarios, giving you a quick readout of best practices and creative ideas for risk management. In fact, AI models (especially those tuned for business or "risk assessment") can provide a quick and structured approach to risk management, helping consultants address potential issues effectively and instill confidence in clients with well-thought-out mitigation strategies. By systematically going through risks with the AI's help, you're less likely to be caught off guard later, and your client will appreciate the diligence.

Risk analysis with AI can also include scenario planning. You might prompt, *"Simulate the impact on our project if [an adverse event] happens (e.g., a key vendor fails to deliver, or a regulatory change hits mid-project). What consequences should we anticipate, and how could we respond?"* The AI can outline a chain of potential effects and even suggest a plan of action. This is akin to having a rehearsal for disasters in a low-stakes environment. While not every scenario it generates will be likely, thinking them through is a valuable exercise that sharpens your readiness. It's worth noting that risk management is a critical aspect of advising clients on strategic decisions, and bringing AI into this process shows that you are leveraging all available tools to safeguard the project's success. Clients often feel reassured when they see a comprehensive risk register and mitigation plan; it signals professionalism and foresight.

Of course, human judgment remains crucial. AI might list generic risks, but you need to filter and prioritize them based on the client's context—some risks might be irrelevant, while others need extra emphasis. And for the mitigations AI suggests, you'll decide which are realistic and who will be responsible for implementing them. Use these prompts as a starting point, then convene with your team or the client's team to discuss and refine. This collaborative review not only improves the plans but also builds buy-in. Everyone knows Murphy's Law—if something can go wrong, it will. By harnessing AI in your risk analysis, you ensure that you've cast the widest net possible in anticipating those "wrongs," and you'll demonstrate adaptability and calm when guiding your client through uncertainties.

Example Prompts for Risk Analysis:

- *"Identify potential risks involved in [project or strategy], categorizing them into areas like financial, operational, technological, and regulatory risks."*

- *"For each key risk identified (e.g., 'Data security breach during the project' or 'Scope creep due to evolving client needs'), provide a brief risk mitigation plan and contingency actions."*

- *"How might changes in external factors (like a sudden market downturn or new legislation) impact our project objectives, and what preemptive measures can we take now to prepare?"*

These prompts help ensure you leave no stone unturned in risk planning. The AI will generate a comprehensive list of possible issues and preliminary ideas on how to tackle them. You can then prioritize these and flesh them out, combining the AI's breadth with your depth of experience to create a robust risk management strategy.

While you're busy delivering on your project and managing risks, it's also wise to keep an eye on the bigger picture of your consulting practice. One way to grow your influence and attract new business is through Thought Leadership—sharing insights and establishing yourself as a go-to expert in your field. Interestingly, the work you do in market research, proposal writing, and risk analysis can feed into thought leadership content. In the next section, we'll examine how AI can help transform your expertise and experiences into compelling articles, posts, or presentations that amplify your voice in the industry.

Thought Leadership

Thought leadership is about articulating and sharing your unique insights with a broader audience, be it through blog posts, articles, white papers, books, podcasts, or speaking engagements. For consultants, establishing thought leadership can differentiate you in a crowded market. It builds credibility (clients trust experts who are published or cited) and can even generate leads as prospects discover your content. However, consistently producing high-quality thought leadership content is time-consuming. This is where AI, and specifically our prompt library, becomes an invaluable partner. You can use AI to brainstorm content ideas, draft outlines, refine your messaging, and even mimic the style of effective communication. The key is to let the AI help you get your ideas on paper faster, so you can focus on injecting your personal expertise and perspective—the authenticity that truly sets thought leaders apart.

One of the first challenges in thought leadership is deciding *what* to write or speak about. Here, AI can assist by surfacing trending topics or persistent pain points in your industry. For example, a prompt like *"What emerging trends in [your industry] are gaining attention and would be valuable to write about for a thought leadership piece?"* will prompt the AI to list areas that are abuzz—perhaps new technologies, regulatory changes, or evolving consumer behaviors. Indeed, identifying emerging trends to showcase your expertise is a smart strategy for thought leadership, and AI's ability to scan a vast corpus of knowledge can point you toward trends you might not have noticed while immersed in client work. Similarly, you could ask, *"What common challenges are [target audience, e.g., CFOs in manufacturing] facing right now that I could address in an article?"* The AI might highlight issues (say, supply chain disruptions or data analytics adoption) around which you can frame a compelling narrative, demonstrating empathy and authority.

Once you have a topic, crafting a compelling narrative is the next step. You might use the AI to generate an outline or even a first draft. A prompt from our library such as *"Outline a thought leadership article on [chosen topic], including an introduction that hooks the reader, three main points with supporting evidence, and a conclusion with a call to action"* will yield a structured game plan for your piece. From there, you can take each section and elaborate. If you get stuck or want to ensure your tone is engaging, try a prompt like *"How can I explain [a complex concept] in a clear, relatable way for [specific audience]?"* The AI's response can give you analogies or simpler wording that make your content more accessible. This is crucial because being a thought leader isn't just about *what* you say, but *how* you say it. Expertise must be coupled with the ability to communicate insights effectively. AI can help translate your deep knowledge into polished, reader-friendly language, essentially acting as an editor or coach pointing out where you might clarify or strengthen your message.

It's worth addressing a subtle point here: can AI itself generate *original* thought leadership? The answer is that AI can generate text that sounds authoritative, but true thought leadership comes from synthesizing information in new ways, sharing personal experiences, or offering provocative viewpoints. AI is great at pulling together known information (what's already been written or said), but the spark of originality comes from you. Think of the AI's contributions as the raw material or the echo of the collective knowledge, which you will mold and refine. For example, the AI might give you a generic paragraph on why a new technology is important, but you should add your own case example or take a stance on its implications. This is how you differentiate through authenticity in your content—by adding stories, opinions, and questions that only *you* might pose. Using the prompts in this library responsibly means not simply accepting whatever the AI says, but critically evaluating it and blending it with your unique perspective. Interestingly, the very act of collaborating with AI can spark your curiosity: it might present a viewpoint that you hadn't considered, to which you can react or rebut in your writing, thereby enriching the final piece.

In practical terms, AI can also assist with the polish. You might take a draft you've written and ask the AI, *"Suggest improvements to the flow and tone of this article. Make it sound engaging and authoritative without being too formal."* It can recommend edits, identify sections that are unclear, or vary the sentence structure to improve readability. It's like having an on-demand editor. But again, you approve which suggestions to take, ensuring the final voice remains yours—human, warm, and insightful. The result is content that is both high-quality and heartfelt. It invites readers to think and demonstrates that you, as a consultant, are always learning and contributing ideas, not just executing projects.

Example Prompts for Thought Leadership:

- *"What are some fresh, pressing topics in the [consulting domain or specific industry] that would resonate with executives and decision-makers right now? Provide a few ideas for thought leadership content."*

- *"Create an outline for a white paper on [topic], including key sections such as Introduction (with a strong thesis), Background, Key Insights or Findings (with sub-points), and Conclusion (with recommendations)."*

- *"Draft a paragraph explaining [complex concept] using a compelling narrative or metaphor that a non-expert audience could easily understand and find intriguing."*

By using prompts like these, you can accelerate the content creation process. The AI will supply the foundation—topic ideas, outlines, even sample text—which you can then build upon with your own knowledge and voice. The end product is thought leadership material that is both informative and authentically you, helping establish your reputation in the marketplace.

Even as you establish your presence through thought leadership, another avenue to engage and provide value is through interactive workshops. Workshops and training sessions allow you to share knowledge in a dynamic, face-to-face (or virtual) environment, often leading to deeper insights and stronger relationships with participants. Designing a high-impact workshop is yet another task where AI prompts can lend a helping hand. Let's delve into how AI can assist in the creative process of workshop design, ensuring your sessions are informative, engaging, and memorable.

Workshop Design

Designing a workshop involves balancing structure and creativity. As a consultant, you might be called upon to run a strategy offsite for a client's leadership team, a training workshop for their staff, or a collaborative problem-solving session with stakeholders. Each scenario requires a tailored approach including specific activities, timings, materials, and facilitation techniques to meet the objectives. It's a task that benefits greatly from both left-brain planning (logistics, sequence, time management) and right-brain innovation (interactive exercises, stories, visual aids). AI can assist with both aspects. With the right prompts, you can generate agendas, brainstorm interactive exercises, and refine the wording of instructions or discussion questions. Particularly for solo consultants or small teams, AI ensures you don't have to reinvent the wheel for every workshop; it's like having a creative co-facilitator in your planning sessions.

Start with the big picture: what do you want participants to walk away with? Suppose you're designing a half-day workshop on "Innovative Thinking for Non-Profit Teams." You could prompt the AI: *"Outline a 4-hour workshop agenda on [topic], including ice-breaker, key activities/exercises, break times, and a wrap-up discussion. Make sure it's interactive and keeps participants engaged."* The AI's response will likely give you a solid structure — perhaps suggesting an opening ice-breaker related to the topic, followed by a mix of short lectures, breakout discussions, hands-on exercises, and a closing reflection. This provides a starting framework that you can adjust based on your knowledge of the client's team (for instance, if they prefer more discussion vs. activities). Consultants have experimented with using AI for such tasks and found that ChatGPT is remarkably capable of generating workshop plans when given suitable prompts. In other words, if you clearly state your workshop goals and parameters, the AI can produce a draft design that hits those marks. This can save you considerable time and spark ideas you might not have come up with solo.

Next, consider the content of each segment. Let's say you want an interactive exercise to teach the concept of risk-taking in innovation. You could ask: *"Suggest a creative group exercise that illustrates the benefits of taking calculated risks in an organization. It should involve role-playing or decision-making under uncertainty."* The AI might propose a simulation or a scenario game; something like a fictitious case where groups must choose between a safe option and an innovative risky option and then debrief on outcomes. Having such raw ideas generated for you helps overcome creative blocks. You can then refine the exercise: tailor the scenario to the client's industry, simplify or elaborate rules, and add your facilitator's touch in how you'll run it. Another example, if you need a visual or story element: *"I need a compelling anecdote or case study to illustrate [concept]—can you provide a short story that I could share with the workshop participants?"* The AI might give you a hypothetical scenario that you can use or adapt to make a point. Throughout this process, maintain your curiosity. Treat the AI as a brainstorming partner throwing suggestions at the wall, some of which will stick and some won't. Even ideas that aren't quite right can lead you to others that are.

One caution: AI might occasionally produce workshop content that sounds good but isn't very practical, or it might lapse into generic corporate language ("project management speak") if not guided properly. Always run a reality check on AI-generated plans. Ask yourself: Will this activity truly engage people? Is the timing realistic? Does the tone fit the audience (e.g., more formal for executives, more playful for junior staff)? You may need to tweak prompts to get the style right—for instance, *"Rephrase the instructions for this exercise in a more casual, encouraging tone."* And as always, inject your authentic style as a facilitator. If you know a certain joke or personal story tends to lighten the mood, plan to include it. If you have read of a successful case from another company that fits an exercise, mention it. These personal elements ensure the workshop doesn't feel canned.

With AI taking care of the foundation and giving you a flow of ideas, you can concentrate on the flow of energy in the workshop: how one segment leads to another, how to keep people energized after lunch, how to handle group dynamics. By the time you deliver the session, you'll have a well-structured agenda (thanks to AI's help) and the confidence that you've thoughtfully tailored it (thanks to your own expertise). The result should be a workshop that not only transfers knowledge but also sparks enthusiasm and empowerment in participants, which is exactly what an AI-enhanced consultant strives for.

Example Prompts for Workshop Design:

- *"Create a draft agenda for a 1-day workshop on [topic]. Include a mix of presentations, interactive group exercises, Q&A sessions, and breaks. The goal is to keep executives engaged and to foster action-able takeaways."*

- *"Give me an idea for an ice-breaker exercise that ties into [work-shop theme]. It should get people talking and thinking about [key concept] from the outset."*

- *"What are some thought-provoking discussion questions I can pose to participants after the segment on [specific subject]? Provide 3-5 questions that will prompt reflection and debate."*

These prompts will help you generate the building blocks of a successful workshop. The AI's output provides structure and creative ideas that you can refine. By combining an AI-generated blueprint with your own facilitation savvy, you'll design workshops that are both methodically sound and uniquely engaging.

After running insightful workshops or completing project deliverables, there often comes a moment when you must present the high-level findings and recommendations to the top brass—the executives or board members who have the final say. These executive briefings demand clarity, brevity, and strategic focus. In the next section, we'll see how AI can help distill complex analyses into crisp, compelling messages suitable for an executive audience, ensuring that your hard-won insights lead to decisions and action.

Executive Briefing

Executive briefings are where all your hard work—the research, analysis, problem-solving, and execution—gets crystallized into a form that busy decision-makers can grasp quickly. In an executive briefing (be it a written report, a slide deck, or a live presentation), less is more. You need to convey the essence of your findings and the rationale for your recommendations without drowning leaders in detail. This is easier said than done, especially when you've been deep in the weeds of a project. AI can serve as a valuable assistant here by helping you summarize information, hone your messaging, and even anticipate the concerns of an executive audience. The prompts in this category of the library focus on refining your communication to be *concise, clear, and outcome-oriented.*

One of the most powerful uses of AI for executive communication is summarization. Suppose you have a lengthy analysis or a detailed technical report that underpins your recommendations. You can prompt the AI: *"Summarize the key insights from [report or analysis details] in a few bullet points suitable for a CEO briefing. Focus on what decisions need to be made and the impact of those decisions."* The AI will attempt to boil it down to the high-level points. You might get bullets like: *"Market demand has shifted by X%, we should pivot product Y accordingly"* or *"Three main risks were identified, of which one is critical to address immediately."* These bullet points give you a starting outline for your briefing. You'll likely need to adjust them, but having a draft summary saves time and ensures you're not missing the forest for the trees. In fact, AI excels at translating complex data into accessible language—exactly what's needed for executive communication. If certain phrasing is too generic, you can refine the prompt: *"Make that summary more succinct and in a confident tone, as if delivering a recommendation to a board."* The iterative prompting will polish the language.

Another angle is tailoring the tone and detail to the audience. Executives typically prefer a broad view with clear options or recommendations, whereas operational teams might want more nuts-and-bolts. AI can help adapt content for different audiences, ensuring you strike the right balance. For example, *"Explain the project results in layman's terms for a non-technical executive, highlighting why these results matter for the business strategy."* This prompt forces the AI to strip away technical jargon and link the outcomes to strategic goals or ROI—which is exactly what the C-suite cares about. Additionally, you could use AI to forecast possible questions or objections an executive might raise. Try a prompt like: *"Given this recommendation (e.g., invest in a new marketing campaign), what questions or concerns might a CEO or CFO have? Provide a list of tough questions."* The AI may produce queries about cost, risk, alignment with strategy, etc. You can use this to prepare answers in advance, effectively rehearsing your meeting. This not only boosts your confidence but also means you come to the briefing with backup slides or data, ready to address those points if they come up.

Ultimately, while AI can help craft the message, delivery is in your hands. The authenticity and confidence with which you speak to executives will influence how your message is received. Use AI's suggestions to ensure your logic is airtight and your words are well-chosen, but when you're in the boardroom (virtual or physical), speak from the heart as well as the mind. Remember that executives are humans too—they respond to narrative and vision. If you can tell a brief story (maybe the "journey" of the project or a customer anecdote that illustrates a key point) that resonates on a human level, do so. You might even use AI to help find that anecdote or analogy, but it's your delivery that gives it life.

In summary, AI-assisted prompting for executive briefings helps you sharpen your key messages and stay concise. By offloading some of the grunt work of summarizing and editing to the AI, you free yourself to focus on strategic insight and emotional intelligence—reading the room, emphasizing the most pertinent points, and connecting your recommendations to the executives' priorities. That's how you ensure your briefings lead to decisions and action, maximizing the impact of your consulting engagement.

Example Prompts for Executive Briefing:

- *"Summarize the findings of our 3-month consulting project into a one-page executive brief. Include: the problem we addressed, the solution implemented, key results (with one statistic if available), and a final recommendation for next steps."*

- *"Convert this detailed analysis into a few high-level talking points for a presentation to the Board. Focus on what decision is needed and why it's important now."*

- *"Anticipate three questions a skeptical CEO might ask about our recommendation to [take some action], and suggest how to answer each convincingly."*

These prompts help distill your work into executive-ready insights. The AI will generate concise summaries and potential Q&A points that ensure you cover what matters most to leaders. You'll still refine and fact-check the outputs, but this assistance can significantly streamline the preparation of briefings that are both informative and impactful.

With the exploration of role-specific prompt use cases complete, it becomes evident that a one-size-fits-all approach to AI prompting isn't sufficient in consulting. The true art lies in customizing prompts for client-specific needs, which we'll discuss next. Just as every client is different, the way you employ AI should adapt to those differences. In the following section, we'll look at how to tailor prompts to fit the unique context of each client and project, ensuring the AI's output is as relevant and useful as possible.

Customizing Prompts for Client-Specific Needs

The prompt examples and strategies we've covered so far provide a versatile toolkit, but to unlock their full potential, you must customize them to your client's specific needs and context. Think of a prompt as a conversation starter with the AI; the more context and nuance you provide, the more tailored and valuable the response will be. In a consulting environment, this means weaving in details about the client's industry, company culture, strategic objectives, and even terminology. A generic prompt might produce generic output, whereas a customized prompt can yield insights that feel like they were written with the client in mind. This is a key differentiator in AI-assisted consulting—your ability to frame the AI's task in a way that mirrors the client's world. In doing so, you ensure that the AI's contributions enhance your originality instead of flattening it.

How do you customize a prompt effectively? First, anchor it in the client's domain. For example, instead of asking *"What are the current trends in supply chain management?"* you could ask *"What current trends in supply chain management are impacting mid-sized retail businesses in [Client's country]?"* By specifying the type of business and region, you guide the AI to filter its vast knowledge for what's most pertinent. If the client has specific language they use (e.g., they refer to "customers" as "members" or have an internal name for a process), include those terms in your prompt. Not only does this tailor the output, but it may also increase the client's acceptance of the recommendations because the language feels familiar. Essentially, you are aligning the AI's output with the client's unique business context and jargon, which can be incredibly powerful for resonance and practicality.

Next, consider the client's objectives and constraints. A prompt for a risk analysis might be very different for a risk-averse client vs. a risk-tolerant one. For the cautious client, you might prompt, *"List potential risks of X initiative for a company that has zero tolerance for regulatory non-compliance."* For the more adventurous client, *"...for a company willing to take bold steps and accept short-term losses for long-term gains."* The AI's answers will likely differ in tone and content, giving you perspective that aligns with the client's risk appetite. This way, you're effectively tuning the AI to operate within the decision framework that the client uses. It's similar to how you would phrase things differently when speaking to a CFO vs. a CTO—you emphasize financial implications for one and technical details for the other. With AI, you explicitly tell it which angle to emphasize.

Another aspect of customization is providing the AI with *contextual data*. For instance, you can feed it a snippet from the client's annual report or a summary of their strategy, and then ask for advice or analysis based on that. A prompt could look like: *"Given the following context about the client's strategic goals [insert a few key points], generate ideas for [the problem at hand]."* By doing this, you're effectively aligning the AI's outputs with the client's known goals and challenges. This leads to recommendations that are not off-the-shelf, but rather interwoven with the client's narrative. Do be mindful of confidentiality and data sensitivity, though—never input sensitive client data into a public AI service. But high-level context or anonymized facts are often enough to steer the AI.

One of the beauties of customizing prompts is that it trains you, the consultant, to clarify what you really need. It encourages a level of precision in thinking: Who is this advice for? What are they trying to achieve? What factors matter most to them? Answering these questions in the form of a prompt not only leads to better AI output, but it also refines your own understanding of the client. It's a reflective exercise. Moreover, by practicing prompt customization, you become faster and more adept at it. Over time, you'll build a mental library of prompt phrasings that worked well for certain industries or scenarios. You might discover, for example, that framing a question in the context of a particular market condition (like *"in a post-COVID world"* or *"in an era of digital disruption"*) yields more relevant insights for clients grappling with those exact issues.

In summary, don't treat the prompts in this book as static recipes. See them as templates to be molded. Tailor every prompt with the client's name, their industry specifics, their values, and their challenges. The extra minutes you spend adding this context can turn a decent AI answer into a game-changing insight. It's the difference between an output that prompts you, the consultant, to say "Okay, that's somewhat helpful," instead of one where you say "Wow, that's exactly what we needed!" Clients may not see this behind-the-scenes work, but they will surely feel the impact in the relevance and quality of your deliverables. Ultimately, customizing prompts is about keeping the client at the center of the AI-enhanced consulting process, which is precisely where they belong.

Tips for Customizing Prompts:

- **Include industry and role details.** Frame your prompt to reflect the client's industry and the role of the advice. For example, *"advise a healthcare clinic"* vs. *"advise a fintech startup"* will yield different nuances.

- **Incorporate client goals.** If the client's goal is, say, *expansion, efficiency,* or *innovation*, mention it. For example, *"Suggest* *marketing strategies for a company focused on sustainable growth in the next 2 years."*

- **Mimic the client's tone.** Adjust the style requested in the prompt to match the client's culture. If they prefer formal reports: *"Provide a formal analysis..."* If they're more informal or visionary: *"Provide an inspiring, big-picture analysis..."*

- **Provide contextual data.** When possible, prepend a short summary of the client's situation. *"Our client (mid-size retailer) has experienced a 10% drop in sales in Q3. Given this context, what...?"* This leads to output that factors in that detail.

- **Iterate and refine.** Don't settle for the first output. If the answer is too generic, refine your prompt with more specifics or instruct the AI to focus on a certain angle. Iteration often produces gold.

By following these tips, you ensure the AI's contributions are not generic platitudes but tailored insights. This customized approach keeps your consulting advice sharp, relevant, and aligned with what matters most to your client.

Having mastered the art of customizing prompts, there is one more frontier to explore: empowering your clients to use AI tools themselves. In the spirit of being a trusted advisor, it's not enough to deliver great work augmented by AI; true empowerment comes when clients can also leverage these tools (under your guidance) to solve problems and innovate. In the final section, we'll discuss strategies for training clients to use AI-enhanced tools, enabling them to carry forward the torch of AI-enhanced productivity and creativity long after your engagement concludes.

Training Clients to Use AI-Enhanced Tools

As consultants, we don't just solve problems—we often teach our clients new ways of thinking and working. In the age of AI-enhanced consulting, one of the most valuable skills you can impart to your clients is the ability to effectively use AI tools themselves. This might seem counterintuitive at first (won't they need us less if they can do it on their own?), but in practice it deepens your role as a trusted partner. By training clients to use AI-enhanced tools, you help them build capacity, which can lead to more so-phisticated collaborations and opens up opportunities for you to tackle higher-level challenges with them. Plus, empowering clients aligns with a human-centered ethos: it shows that you prioritize their long-term success over short-term dependencies. The tone here should be encouraging and con-fidence-building; many clients may feel intimidated by AI, and it's our job to demystify it and make it accessible.

When introducing clients to AI tools (like a chatbot for research or a custom AI application for their business), start with the basics. Show them some of the very prompts from this library and the results these prompts can generate. A great opening exercise is to pick a simple, relevant problem and walk through it together using an AI assistant. For instance, if your client is a small business owner struggling with social media content, demonstrate a prompt like *"Generate five ideas for Instagram posts that a [type of business] can use to engage customers."* As the AI produces suggestions, the client often lights up; seeing possibilities unfold in seconds is a powerful moment. How-ever, equally important is teaching them to not take the outputs at face value. Emphasize that AI-generated answers should be taken with a grain of salt and reviewed critically, just as they would any advice. This helps manage expectations: the tool accelerates thinking, but human judgment must vali-date the results.

A simple framework to share with clients is the "Guide, Don't Abdicate" principle. Encourage them to treat AI as a smart colleague: you guide it with clear instructions, and it gives you a draft or an idea, which you then refine. For example, train them in refining prompts. If the first answer is too broad or misses the mark, show them how to tweak the question to get closer to what they need (much as we've done in previous sections). Maybe the output was too generic—suggest adding a detail to the prompt. Maybe it was off-topic—explain how to explicitly state the context next time. This iterative dialogue is a skill in itself. In fact, one tip is to continue refining prompts to generate more accurate responses faster. Clients should understand that prompt crafting is an interactive process, not a one-shot deal. You might role-play with them: have them pretend to be the AI and you be the user, or vice versa, to illustrate how specificity and clarity in prompts lead to better outcomes.

Another key lesson for clients is the importance of custom instructions or settings in AI tools. Some platforms allow users to set a profile or preferences (for instance, "assume a formal tone," or "my business is X, tailor responses accordingly"). Show them how to use these features if available, as they can drastically improve the relevance of outputs. This is akin to configuring a tool correctly before use. If you train an internal team, for example, you might help them set up an AI knowledge base with their own data, so that the AI's answers are grounded in their reality. This way, the AI becomes more than just a clever toy, it becomes an integrated part of their workflow.

Crucially, address the fears and ethical considerations. Clients may worry: "Will AI make mistakes that embarrass us? Could it divulge sensitive info? Will it replace jobs?" Speak to these honestly. Acknowledge that AI can make factual errors (hallucinations), so they must fact-check important outputs. Advise on boundaries. For instance, *never paste confidential or personally identifiable information into a public AI service* to avoid leaks. Encourage them to start with low-risk uses of AI and gradually increase as they gain confidence. On the job replacement front, emphasize that AI is a tool to augment human capability, not a replacement for human creativity and decision-making. By adopting it early, they're actually making their team more valuable by freeing humans to focus on higher-order tasks. In our experience, framing AI as assisting with the drudgery (summarizing documents, drafting routine communication, etc.) helps people see it as a colleague rather than a threat.

Finally, share success stories (anonymized if needed) of other clients or firms that adopted AI in their operations successfully. For instance, mention how a client's marketing team cut their content creation time in half by using AI for first drafts, or how a strategy team used AI to model scenarios that informed a major decision. These stories serve to inspire and make the potential tangible. They also reinforce a culture of curiosity and continuous learning, showing that experimenting with AI can yield positive surprises and breakthrough ideas. In training sessions, keep the atmosphere light and experimental. Encourage clients to play with prompts, even silly ones, to get comfortable. The more they play, the more creative they become in applying the tool.

By the end of your training, the ideal outcome is a client team that feels *empowered and excited* to use AI in their daily work. They'll view you as a consultant who can deliver a project today who has also upgraded their capabilities for the future. This positions you as a long-term partner in their success. And as they use these tools, they may uncover new questions or needs—naturally turning back to you for deeper consulting engagements. Thus, in teaching them, you're also sowing seeds for future collaboration built on trust and shared progress.

Key Points to Emphasize When Training Clients:

- **Start simple.** Begin with easy, relevant use cases where AI can help (e.g., brainstorming ideas, drafting an email) to build confidence.

- **Critical evaluation.** Always review AI outputs; encourage a mindset of "trust but verify." Mistakes can happen, so double-check important facts or calculations.

- **Iterative prompting.** Demonstrate how refining a question can change the answer. Show them that asking follow-up questions or adding detail leads to better results.

- **Privacy and ethics.** Instruct what not to share with public AI tools (sensitive data) and discuss responsible use. Set guidelines so they feel safe and informed.

- **Practice and support.** Provide cheat sheets or a few example prompts (perhaps drawn from this prompt library) as a starting point. Encourage practice and let them know they can reach out with questions as they integrate AI into their work.

By covering these points, you prepare your clients to incorporate AI tools into their operations wisely and effectively. This not only adds value to the engagement but also elevates the client's own capabilities. In turn, you strengthen your reputation as a forward-thinking, client-centered consultant who empowers others.

In closing, the prompts and narratives presented in this appendix underscore a central theme: AI is a powerful enabler for consultants, but it is most effective when guided by human insight, creativity, and integrity. Whether you're diving into market research, crafting proposals, onboarding clients, analyzing risks, authoring thought leadership, designing workshops, or briefing executives, there's a prompt (likely several) that can accelerate and enhance your work. Use this prompt library as a starting point, and don't hesitate to modify each prompt to fit your voice and your client's unique situation.

Remember that the AI-enhanced consultant is not a consultant who offloads thinking to machines, but one who uses these tools to think bigger and deliver more. The tone we set—empowering, curious, authentic—ensures that we leverage AI in a way that differentiates our work rather than homogenizes it. By customizing prompts and training clients, we ensure that our use of AI remains human-centric: solving real problems for real people, and expanding what's possible for those we serve.

As you integrate these practices into your consulting work, you'll likely develop your own library of favorite prompts and techniques. Embrace that evolution. Stay curious about new AI capabilities that emerge, and continue to refine your approach in alignment with your values and your clients' best interests. In doing so, you'll not only stay relevant in the rapidly changing landscape—you'll define it, as a pioneer at the intersection of human expertise and artificial intelligence.

ROB BERG

◆

Glossary of AI Terms

A

Agentic Workflow (Autonomous AI Agents). Refers to processes where AI agents operate with some autonomy, chaining tasks and using tools to achieve goals with minimal human intervention. This concept, demonstrated by systems like AutoGPT, is relevant to consultants exploring how to automate complex multi-step workflows using AI.

AI Analytics. Analysis that uses machine learning to process large amounts of data to identify patterns, trends, and relationships without requiring human input. For consultants, AI analytics tools can uncover business insights from big data faster and augment human analysis capabilities.

AI Assistant. A program (often a chatbot or virtual assistant) that uses AI to understand and respond to human requests, schedule meetings, answer questions, and automate repetitive tasks. Professionals leverage AI assistants to offload routine tasks and improve response times in client communications.

AI Bias. The tendency of AI systems to produce outputs that reflect or amplify biases present in their training data, potentially reinforcing harmful stereotypes. Consultants must be aware of AI bias to ensure the solutions they develop are fair and do not unintentionally discriminate or skew results.

AI Change Management. The process of effectively managing the organizational changes that come as a result of adopting, integrating, and implementing AI in the workplace. Consultants guide businesses through AI change management to ensure employees embrace new AI-driven work-flows and that transformations yield lasting benefits.

AI Ethics. The field concerned with ensuring AI is developed and used in ways that align with human values, prevent harm, and address the moral questions raised by AI technology. In consulting, AI ethics translates to creating governance frameworks so that AI initiatives are responsible and trust-building.

AI Governance. Refers to the policies, regulations, and frameworks established to ensure the ethical and responsible development, deployment, and use of AI within an organization. Consultants help design AI governance structures to manage risk and compliance in enterprise AI projects.

Algorithm. A formula or set of rules that a computer follows to solve problems or make calculations. In AI consulting, understanding algorithms is foundational, as they underpin machine learning models and decision systems built for clients.

Artificial General Intelligence (AGI). A hypothetical type of AI that would possess the ability to understand, learn, and apply knowledge across a wide range of tasks, much like a human. AGI remains theoretical; consultants discuss AGI mainly in terms of future possibilities and long-term strategy rather than current practical use.

Artificial Intelligence (AI). Technology that enables machines to simulate human cognitive functions such as learning, problem-solving, and decision-making. In business, AI powers data analysis, personalization, and predictive modeling, making it a powerful tool for consultants driving digital transformation.

Artificial Superintelligence (ASI). A hypothetical future stage of AI development where AI systems would surpass human intelligence and capabilities. While ASI is speculative, consultants might reference it in thought leadership on the ultimate potential and risks of AI.

Augmented Intelligence. A design approach to AI emphasizing a human–AI partnership, where AI systems enhance human cognitive performance rather than replace it. Consultants often frame AI projects as augmented intelligence to focus on empowering employees with AI tools for better decision-making.

AutoGPT. An experimental open-source project that chains multiple AI model prompts to attempt multi-step tasks autonomously. AutoGPT exemplifies an "AI agent" that iteratively plans and executes actions towards a goal, a concept consultants watch closely for its potential to automate complex tasks with minimal oversight.

Autonomous Decision-Making Systems. AI tools that make operational decisions without human intervention, such as programmatic advertising platforms that automatically adjust bids based on real-time data. Consultants evaluate these systems to automate client business processes while ensuring appropriate safeguards and controls.

Autonomous Machine. A device or system that can learn, reason, and make decisions using available data without human intervention. Self-driving vehicles are a prime example; consultants might work on assessing the readiness of such machines for client operations or integrating them into business models.

B

Big Data. Extremely large datasets that can be analyzed computationally to reveal patterns, trends, and associations. Handling Big Data is central to AI projects—consultants help organizations manage and extract value from Big Data to train models and inform strategy.

C

Chatbot. A software application designed to conduct conversations with human users, often used for customer service, information retrieval, or task automation. Consultants deploy AI chatbots to help clients improve customer engagement and automate routine Q&A through natural language interaction.

Client Onboarding Automation. The use of AI and automation tools to streamline the process of welcoming and integrating new clients. Client onboarding automation minimizes manual work required to register a new client and set up services. Consulting firms implement such automation to improve client experience, accelerate onboarding timelines, and reduce errors in the intake process.

Cloud Computing. The delivery of computing services (servers, storage, databases, software, etc.) over the internet (the "cloud"). Cloud resources are used to power many AI applications and services, allowing consultants to build scalable, on-demand AI solutions without managing on-premise infrastructure.

Computer Vision. An AI technology that enables machines to interpret and understand visual information from the world. Business applications of computer vision include inventory management, quality control, and customer behavior analysis in stores, which consultants leverage to drive operational efficiencies for clients.

Conversational AI. Technologies that enable machines to engage in human-like dialogue, processing and responding to natural language input. This encompasses AI-driven chatbots and virtual assistants that consultants implement for clients to handle customer inquiries, internal helpdesk requests, and other interactive communications.

D

Data Mining. The process of discovering patterns, correlations, and insights in large datasets using statistical, machine learning, and computational techniques. In consulting, data mining is used to extract valuable business intelligence (e.g., customer trends or process inefficiencies) that can inform decision-making.

Data Science. An interdisciplinary field that uses statistics, computer science, and domain knowledge to extract insights and predictions from data. For consultants, data science drives the development of predictive models and analytics solutions that solve client problems and unlock new opportunities from data.

Deep Learning. A specialized subset of machine learning that uses artificial neural networks with many layers (hence "deep") to model complex patterns in data. Deep learning excels at tasks like image recognition, speech recognition, and natural language processing, enabling consultants to tackle complex problems (such as vision or language tasks) with high accuracy.

E

Edge AI. AI processing that occurs locally on devices at the "edge" of the network, rather than in the cloud. Edge AI enables faster responses and greater privacy (since sensitive data can be processed on-device), which consultants consider for solutions like real-time analytics in IoT devices or on-site predictive maintenance.

Edge Computing. A distributed computing framework that brings computation and data storage closer to data sources (e.g., IoT sensors). By reducing latency and bandwidth usage, edge computing complements AI by allowing consultants to deploy intelligent services in remote or bandwidth-constrained environments (such as factories or retail stores).

Embedding (Vector Embedding). A numeric vector representation of a data point (e.g., a word, sentence, or image) that captures its semantic meaning. Embeddings convert unstructured data into an array of numbers that still expresses the data's original context. Consultants use embeddings in AI solutions to enable semantic search and recommendations, comparing items by meaning rather than exact keywords.

Explainable AI (XAI). AI systems designed to make their functioning transparent and interpretable to humans. Explainable AI is crucial in consulting engagements because clients need to understand how an AI model arrives at its decisions—providing this transparency builds trust and aids in compliance with regulations.

F

Feature Engineering. The process of selecting and transforming the most relevant variables from raw data to improve machine learning model performance. In consulting projects, data scientists perform feature engineering to ensure models capture the key signals in client data that drive accurate predictions.

Feature Extraction. A technique of identifying distinctive attributes within data inputs that machines can use to classify and understand information. For example, in image data a feature extraction method might detect edges or shapes. Consultants use automated feature extraction to let AI models learn important patterns without manual selection, which accelerates model development.

G

Generative AI. AI systems that create new content by learning patterns from existing data. These systems can produce text, images, audio, video, code, or synthetic data based on their training. Consultants explore generative AI to help clients with content creation tasks, such as generating marketing copy or prototypes, and to augment creative workflows.

Generative Adversarial Networks (GANs). A machine learning framework where two neural networks (a generator and a discriminator) compete against each other, resulting in the creation of increasingly realistic synthetic content. Consultants might use GANs to generate synthetic data for training models or to create simulations for scenario planning.

GPT (Generative Pre-trained Transformer). An OpenAI-developed large language model architecture that is pre-trained on vast amounts of text data. GPT models (e.g., GPT-4o) understand natural language inputs and can generate human-like text for various applications, which consultants integrate into solutions like AI chatbots, report generation tools, or coding assistants.

H

Hallucination. When AI systems, particularly large language models, produce factually incorrect information that appears plausible but has no basis in the data or reality. Consultants need to mitigate AI hallucinations (for instance, by using Retrieval-Augmented Generation or human review) to ensure the AI solutions they deliver provide reliable and accurate outputs.

Human-in-the-Loop (HITL). Systems where human feedback and interaction is integrated into the training, operation, or refinement of AI models. Incorporating a human-in-the-loop (for example, to review AI decisions or label data) is a strategy consultants use to improve model accuracy, inject domain expertise, and maintain oversight on critical AI-driven processes.

I

Image Recognition. The ability of AI systems to identify objects, people, places, or actions in images or video content. Also called image classification, this technology powers visual search, content moderation, and automated tagging. Consultants apply image recognition in use cases like detecting defects in manufacturing or analyzing customer behavior from security footage.

Insight-as-a-Service. The delivery of actionable analytical insights to organizations through a service-based model. Rather than just providing software or raw data, consultants offering Insight-as-a-Service use AI and analytics to continuously generate valuable findings (e.g., trend reports, forecasts) for clients, effectively packaging data-driven insights as a managed service.

L

LangChain. An open-source framework for building applications that chain together multiple LLM operations and tools. LangChain is designed to simplify developing workflows for large language models by allowing intermediate steps like data retrieval or transformations between prompts. Consultants use LangChain to rapidly prototype AI solutions (such as custom chatbots or research agents) that require an LLM to interact with external data sources and services.

Large Language Model (LLM). A deep-learning model trained on massive text datasets that can understand, generate, and manipulate human language. LLMs (like GPT-3, GPT-4, etc.) underpin many modern AI applications—consultants leverage LLMs to build advanced conversational agents, summarization tools, and other NLP-driven solutions for clients.

Low-Code Development Platform. A software platform that enables the creation of applications through visual interfaces and minimal hand-coding. Low-code platforms let consultants and developers build AI-enhanced solutions faster by dragging and dropping components (for example, integrating an AI API into a dashboard) without writing extensive code, accelerating delivery of business apps.

M

Machine Learning (ML). A subset of AI where systems improve performance through experience without explicit programming, by learning patterns from historical data. In consulting, ML techniques are used to develop predictive models (like demand forecasting or customer segmentation) that evolve as they process more client data, enabling data-driven decision making.

Machine Learning Operations (MLOps). The practice of managing the end-to-end life cycle of machine learning models, from development and deployment to monitoring and maintenance. MLOps combines principles from DevOps and data engineering to ensure that AI models in production are reliable, scalable, and continuously delivering value. Consultants establish MLOps processes for clients to smoothly deploy AI models and keep them performing well over time.

Model Context Protocol (MCP). A structured architecture that enables large language models to perform multi-step tasks by defining context windows, tool access, memory structures, and agent interactions. MCP coordinates how a model accesses relevant knowledge and executes reasoning across tools, making it useful for consultants building AI assistants that require persistent memory, modular capabilities, or agent collaboration.

N

Narrow AI (Weak AI). AI systems designed to perform specific tasks, but which are not able to function beyond their limited domain. Most current AI applications (e.g., a model that only classifies images or only recommends movies) are forms of narrow AI. Consultants focus on narrow AI solutions that solve particular business problems, while communicating that these systems don't possess general intelligence.

Natural Language Generation (NLG). A subfield of AI that focuses on generating coherent, contextually relevant human language text from structured data. Consultants implement NLG for use cases like automated report writing or personalized client communications, where the AI can turn data (e.g., financial numbers or analytics results) into readable narratives.

Natural Language Processing (NLP). A field of AI that focuses on the interaction between computers and humans through natural language. It enables machines to interpret and generate human language, encompassing tasks such as text analysis, translation, sentiment analysis, and speech recognition. NLP is fundamental in consulting projects involving text data or conversational interfaces.

Neural Networks. Computational models inspired by the human brain's network of neurons, consisting of interconnected nodes that process data in layers. Neural networks are fundamental to deep learning and are used for tasks such as image recognition, speech processing, and language understanding. Consultants often choose neural network architectures as the engine for AI solutions that require learning complex patterns from examples.

Neuromorphic Computing. Computing architecture designed to mimic the structure and function of the human brain, potentially enabling more efficient AI processing. While still experimental, neuromorphic computing could allow consultants to deploy AI models with significantly improved energy efficiency and speed for specialized applications in the future.

No-Code Development Platform. A platform that allows building applications through graphical user interfaces and configuration without writing any code. These tools empower consultants and business users to integrate AI capabilities or automate processes without programming expertise—for example, creating a workflow that uses an AI service (like sentiment analysis) via simple configuration instead of custom code.

O

Overfitting. Occurs when a machine learning model learns the training data too well—including noise or random fluctuations—resulting in poor generalization to new data. An overfit model performs excellently on training data but poorly on unseen data. Consultants watch out for overfitting during model development, employing techniques like cross-validation or regularization to ensure models will be reliable in real-world use.

P

Pinecone. A managed vector database platform designed to store and query high-dimensional vector embeddings efficiently. Pinecone enables developers to perform fast similarity searches through billions of vector data points. Consultants leverage Pinecone to add semantic search and memory capabilities to AI solutions (as part of Retrieval-Augmented Generation pipelines), allowing LLMs and other AI systems to quickly retrieve relevant information from a client's knowledge base.

Predictive Analytics. Involves using statistical algorithms and machine learning techniques to analyze historical data and make predictions about future events. Predictive analytics is commonly applied in fields like finance (e.g., credit risk scoring), marketing (e.g., customer churn prediction), and operations (e.g., demand forecasting). Consultants develop predictive analytics models to help clients anticipate trends and inform strategic decisions.

Prompt Engineering. The practice of designing and refining prompts to elicit the desired responses from AI systems, especially language models. Prompt engineering is crucial for improving the performance and accuracy of AI applications like chatbots or writing assistants. Consultants skilled in prompt engineering can get more reliable and relevant outputs from AI models for their clients, without changing the underlying model.

R

Reactive Machines. The simplest form of AI systems that can only react to current situations without memory of past events or the ability to learn from experience. These systems respond to stimuli in real-time (for example, a chess AI that calculates the best move given the current board state each turn). Consultants may reference reactive machines as one category in the taxonomy of AI capabilities when educating clients on AI's evolutionary stages.

Reinforcement Learning (RL). A machine learning approach where an agent learns to make decisions by taking actions in an environment and receiving rewards or penalties. Over time, the agent learns an optimal strategy (policy) for maximizing cumulative reward. Consultants apply reinforcement learning in scenarios like recommendation systems or robotic process optimization, where an AI can iteratively learn the best actions through trial and feedback.

Responsible AI. The practice of designing, developing, and deploying AI systems in ways that are ethical, transparent, accountable, and aligned with human values. Responsible AI principles guide consultants to consider factors like fairness, interpretability, privacy, and governance in every AI project, ensuring the AI solutions not only achieve business goals but also adhere to societal and regulatory expectations.

Retrieval-Augmented Generation (RAG). An AI technique where a language model's outputs are augmented by retrieving relevant information from a knowledge base or documents and providing it as context. In practice, RAG pipelines fetch facts from sources (often via a vector database) before the model generates an answer, thereby grounding the response in real data. Consultants design RAG-based solutions so that AI assistants and chatbots can give accurate, up-to-date answers using client-specific information rather than relying solely on the model's training data.

Robotic Process Automation (RPA). Software technology that automates routine, rule-based digital tasks by creating "bots" that mimic human actions. RPA can significantly reduce manual effort and operational costs (by 25–40% in some cases) by handling repetitive processes in finance, HR, customer service, and more. Consultants integrate AI with RPA (for example, using AI for decision steps and RPA for execution) to help clients hyper-automate workflows for efficiency gains.

S

Semantic Analysis. The process by which AI systems interpret meaning from language, accounting for context, intent, and nuance beyond simple keyword matching. Semantic analysis techniques enable applications like sentiment analysis or topic extraction. In consulting, these techniques are used to help clients glean insights from unstructured text data (e.g., analyzing customer feedback for themes and sentiments).

Semantic Search. An information retrieval approach that uses the meaning and context of queries and documents, rather than relying on exact keyword matches. By leveraging embeddings to capture query intent and content semantics, semantic search finds results that are contextually relevant. Consultants implement semantic search so that clients can query large text corpora (like document archives or knowledge bases) and get results based on conceptual relevance, improving discovery of insights.

Sentiment Analysis. An AI technique that identifies and categorizes opinions expressed in text to determine the emotional tone (positive, negative, neutral). Often applied to social media, reviews, or customer feedback, sentiment analysis helps consultants assess public or customer sentiment at scale, enabling clients to measure brand perception or the impact of initiatives in real time.

Supervised Learning. A machine learning approach where algorithms are trained on labeled data with known inputs and outputs, allowing them to learn a mapping and then predict outcomes for new, unseen data. Most common AI models in business (classification or regression tasks) use supervised learning. Consultants prepare training datasets and supervise the training process to build accurate models for predictions like fraud detection or product recommendation.

Synthetic Data Generation. The creation of artificial datasets that mimic real information, used to train AI systems without using sensitive or proprietary real data. Consultants may employ synthetic data to augment limited datasets or to protect privacy—for example, generating realistic synthetic customer records to train a model when actual data is scarce or confidential.

T

Token. The basic unit of text that language models process, which can be a word, part of a word, or even just a character depending on the model's design. Tokens are the elements counted for model input/output length (and often for pricing of AI services). Understanding tokens helps consultants optimize prompt sizes and manage costs when using large language models.

Training Data. The dataset used to teach machine learning models, consisting of input examples and their expected outputs. High-quality training data is critical. Consultants spend significant effort ensuring the training data for AI projects is relevant, clean, and sufficiently representative of the problem so the model can learn effectively.

Transfer Learning. A machine learning technique where knowledge gained from training a model on one task is transferred to a different but related task. This approach reduces the amount of new data and training time needed. In practice, consultants use transfer learning by taking pre-trained models (for example, an image recognition model trained on a large dataset) and fine-tuning them on a client's specific data, accelerating development and improving performance with limited data.

Turing Test. A method proposed by Alan Turing in 1950 to determine if a machine exhibits intelligent behavior indistinguishable from that of a human. In a classic Turing Test, a human evaluator interacts with an unknown interlocutor (machine or human) via text; if the evaluator cannot reliably tell which is which, the machine is said to have passed the test. Consultants reference the Turing Test when discussing AI capabilities and human-like performance, and the fact that modern AI (like chatbots) have come close to this benchmark in narrow contexts.

U

Underfitting. Occurs when a machine learning model is too simple to capture the underlying patterns in the data. An underfit model performs poorly on both training and new data (failing to learn the signal at all). Consultants recognize underfitting as a sign that a more complex model or better features may be needed, ensuring the final model is sufficiently expressive to handle the client's problem.

Unsupervised Learning. A machine learning approach where algorithms analyze unlabeled data to find patterns or groupings without explicit guidance. Unsupervised learning techniques (like clustering or anomaly detection) are used by consultants for exploratory data analysis—for example, segmenting customers into unknown groups based on behavior, which can reveal insights without pre-defined labels.

V

Vector Database. A specialized database optimized for storing and querying high-dimensional vector embeddings, primarily used to enable fast semantic search in RAG systems. Vector databases index vectors so that similarity searches (finding nearest neighbors in vector space) are efficient at scale. Consultants incorporate vector databases (such as Pinecone, Weaviate, or FAISS) when building AI solutions that require retrieving information based on semantic similarity (e.g., finding related documents or images).

Virtual Assistants. AI-powered interfaces (often voice- or text-based) that handle user inquiries and tasks through conversational interactions. Operating 24/7, virtual assistants like Siri, Alexa, or custom chatbots can provide customer support, schedule appointments, or answer FAQs without human intervention. In consulting, deploying virtual assistants is a way to improve client customer service operations or employee helpdesks with scalable AI support.

Z

Zero-Shot Learning (ZSL). Zero-shot learning is an AI capability where a model can correctly perform a task or classify information without having been explicitly trained on examples of that task. Instead, it relies on general knowledge, embeddings, and natural language understanding to generalize from related data. A consultant could apply zero-shot learning to accelerate business problem-solving without the need for costly, custom datasets.

Acknowledgements

M ost books owe their existence to a long list of contributors—publishers, editors, reviewers, proofreaders, designers, and countless others who help bring a manuscript to life. This book is different. The work you hold in your hands was created by me, in collaboration not with a traditional publishing team, but with ChatGPT.

That partnership may sound unusual, but it reflects the very subject of this book: the integration of artificial intelligence into professional practice. With ChatGPT, I had at my side not only a capable writing partner but also, indirectly, the collective wisdom of countless experts, researchers, and practitioners whose insights form the backbone of its large language models. Every sentence generated, every idea expanded, and every reference surfaced was shaped by that vast body of human knowledge.

So while no proofreader or editorial board was consulted, I was not alone in this process. I am indebted to the unseen community of scientists, engineers, and creators who built the models, and to the millions of writers, thinkers, and subject-matter experts whose published works provided the foundation on which this AI was trained. They may never know how their contributions helped inform this book, but I recognize and appreciate their influence.

To those millions of contributors, and to the teams who continue to refine and expand the possibilities of AI, thank you. This book is as much a reflection of your work as it is mine.

And to you, the reader, thank you for taking this journey with me. My hope is that, in some small way, these words—shaped by both human and machine—help you find new ways to shape your own work in the age of AI.

ROB BERG

About the Author

R ob Berg has been called "a consultant's consultant." With a career spanning more than thirty years, he has helped build and lead multiple seven-figure consulting practices and has guided countless professionals in elevating their craft. His previous book, *The Courageous Consultant*, has been touted as "required reading for any consultant who wants to be the best in their field" by Andrew Neitlich, Director of the Center for Executive Coaching, and "an inspiring and down-to-earth guide to building a successful and satisfying consulting practice" by Kirkus Reviews.

Rob currently serves as a principal at Perr&Knight, one of the nation's leading niche consulting firms, where he founded their Operations & Technology Consulting practice and consults internally on practical applications of AI. His work has shaped how tech-forward companies, financial services firms, and professional service organizations adapt to technological change while maintaining the human qualities that set them apart.

As an executive coach, Rob works one-on-one with consulting professionals and technology executives to help them reach the top tier of their profession. Known for his ability to blend insight, empathy, and candor, he equips his clients to achieve lasting impact in their work.

Beyond consulting and coaching, Rob is a lifelong learner with passions for photography, astronomy, and jazz. He lives in St Augustine, Florida with his wife Christine and their dogs Jasper and Rupert, and continues to dedicate his time to inspiring others to express themselves creatively while helping to reshape organizations in a time of profound change.

www.ingramcontent.com/pod-product-compliance
Lightning Source LLC
Chambersburg PA
CBHW061203220326
41597CB00015BA/1328